# blueberries

## 50 TRIED & TRUE RECIPES

Julia Rutland

Adventure Publications
Cambridge, Minnesota

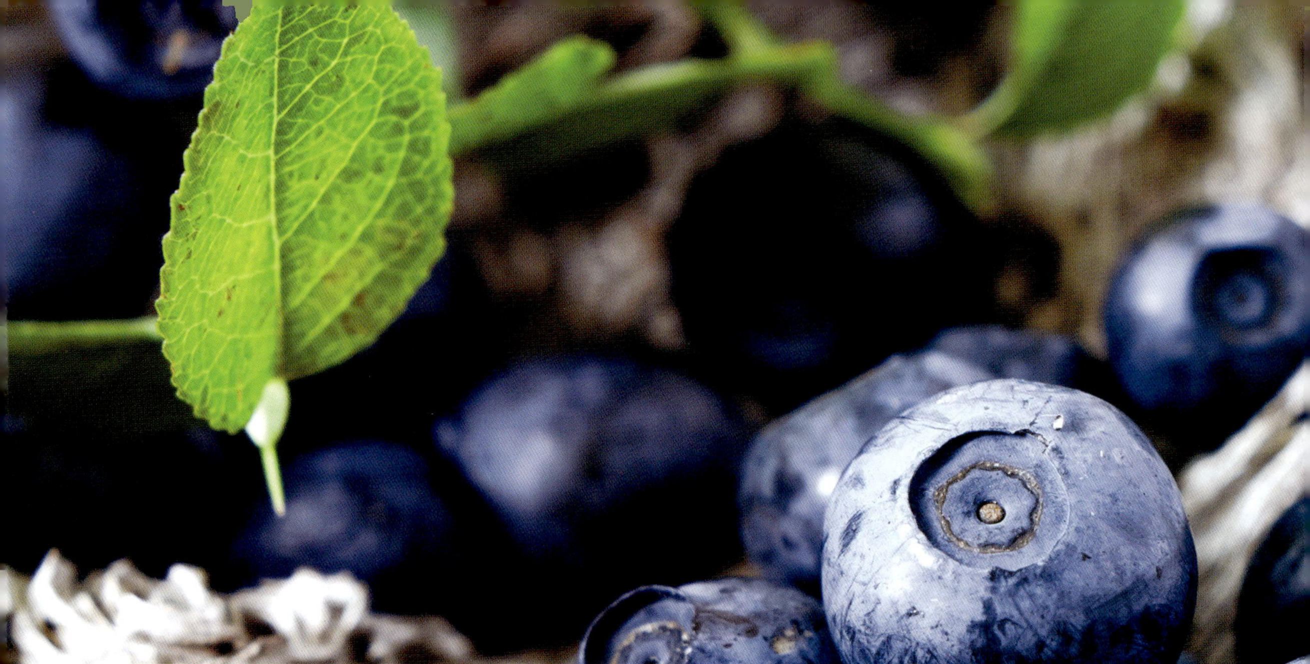

Cover and book design by Lora Westberg
Edited by Emily Beaumont


Cover images: All images by Julia Rutland unless otherwise noted.
**Africa Studio/shutterstock.com:** cone, back cover; **baibaz /shutterstock.com:** jam; **Prolightstudio/shutterstock.com:** background; **Yellow Cat/shutterstock.com:** cookie

All images by Julia Rutland unless otherwise noted.
Used under license from Shutterstock.com:
**Sunny Forest:** 2-3; **HandmadePictures:** 17; **J.chizhe:** 4-5; **valzan:** 9

10 9 8 7 6 5

**Blueberries: 50 Tried & True Recipes**

Published by Adventure Publications
An imprint of AdventureKEEN
310 Garfield Street South
Cambridge, Minnesota 55008
(800) 678-7006
www.adventurepublications.net

Printed in China
ISBN 978-1-59193-847-7 (pbk.); 978-1-59193-848-4 (ebook)

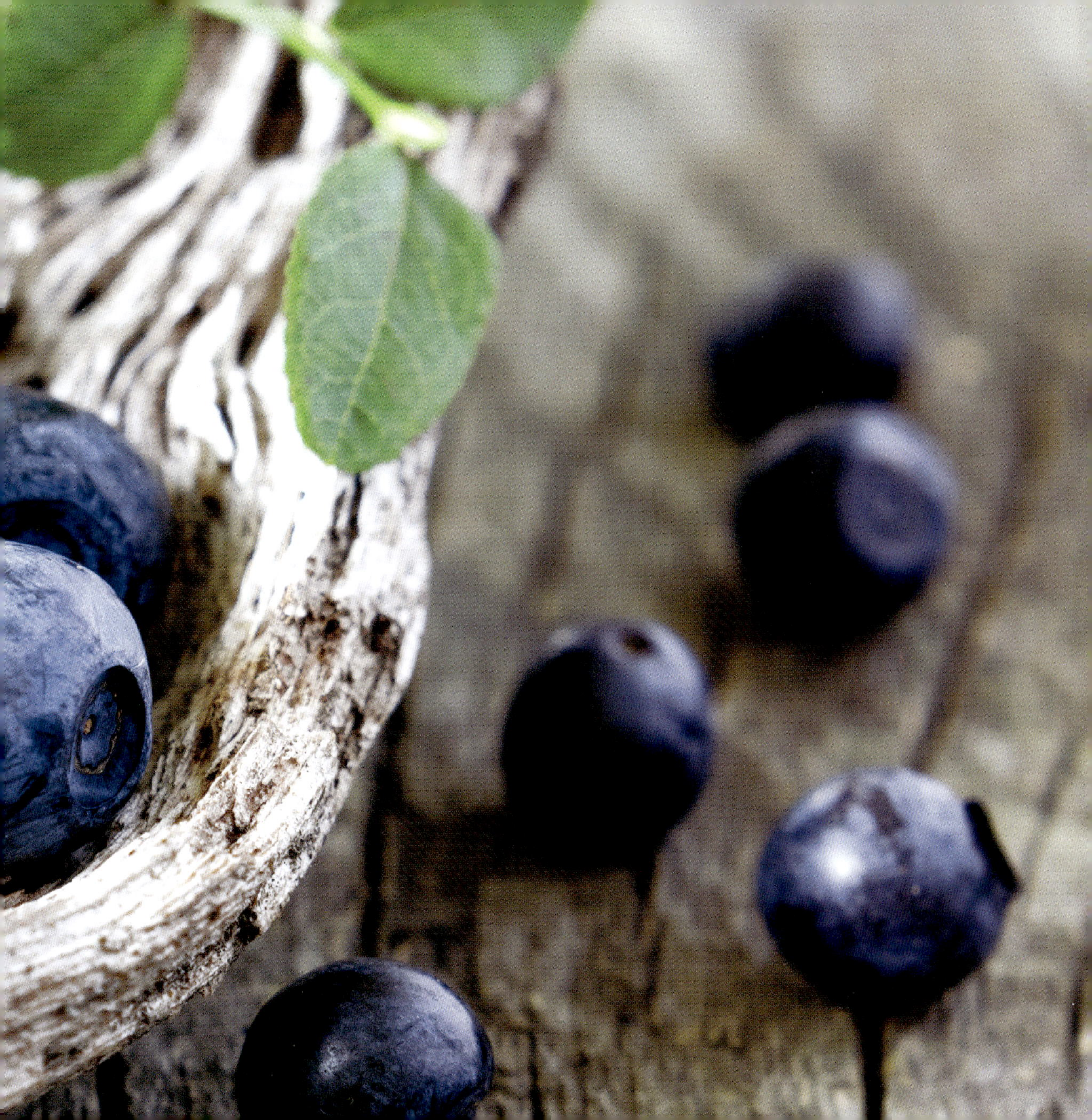

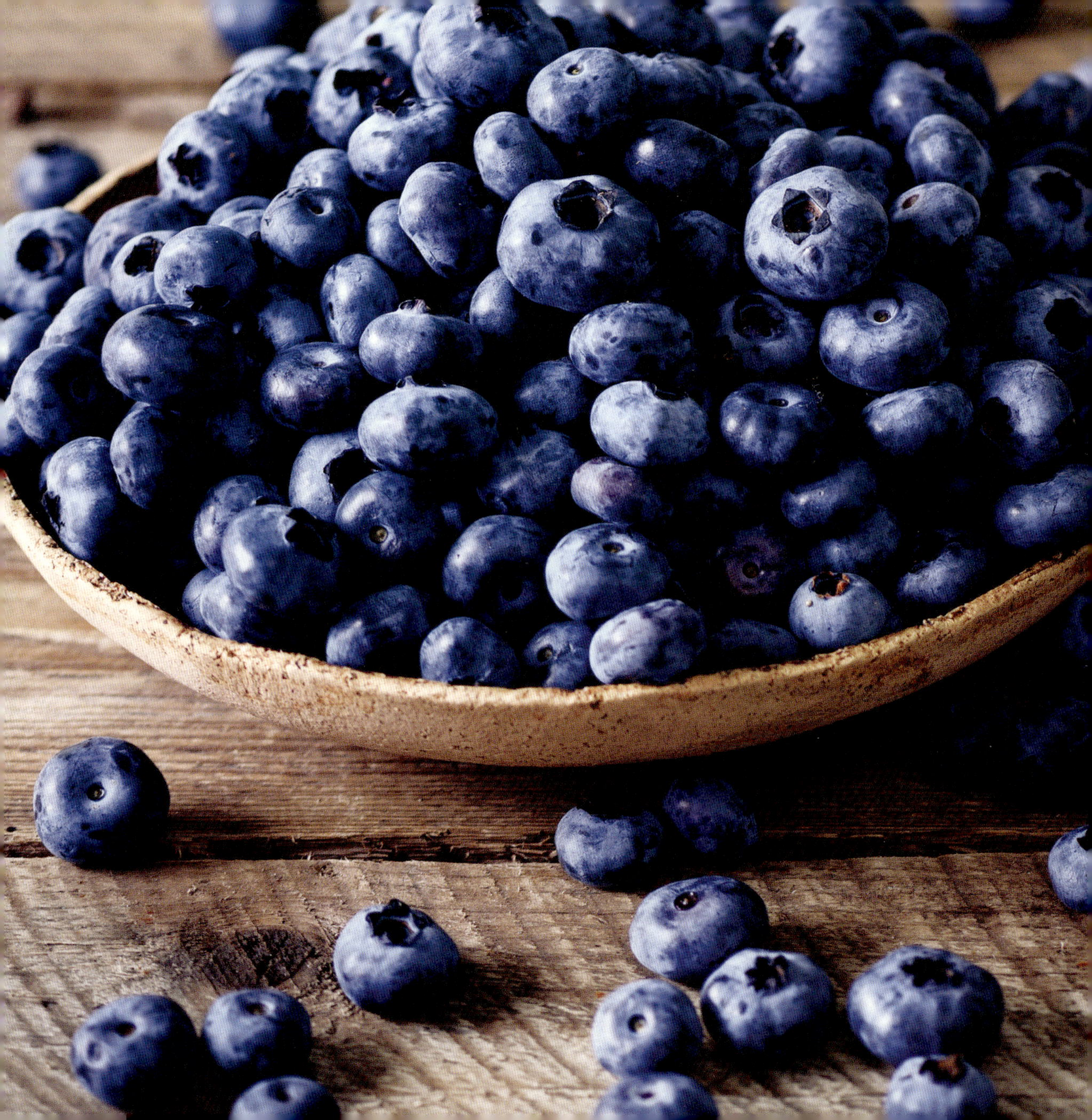

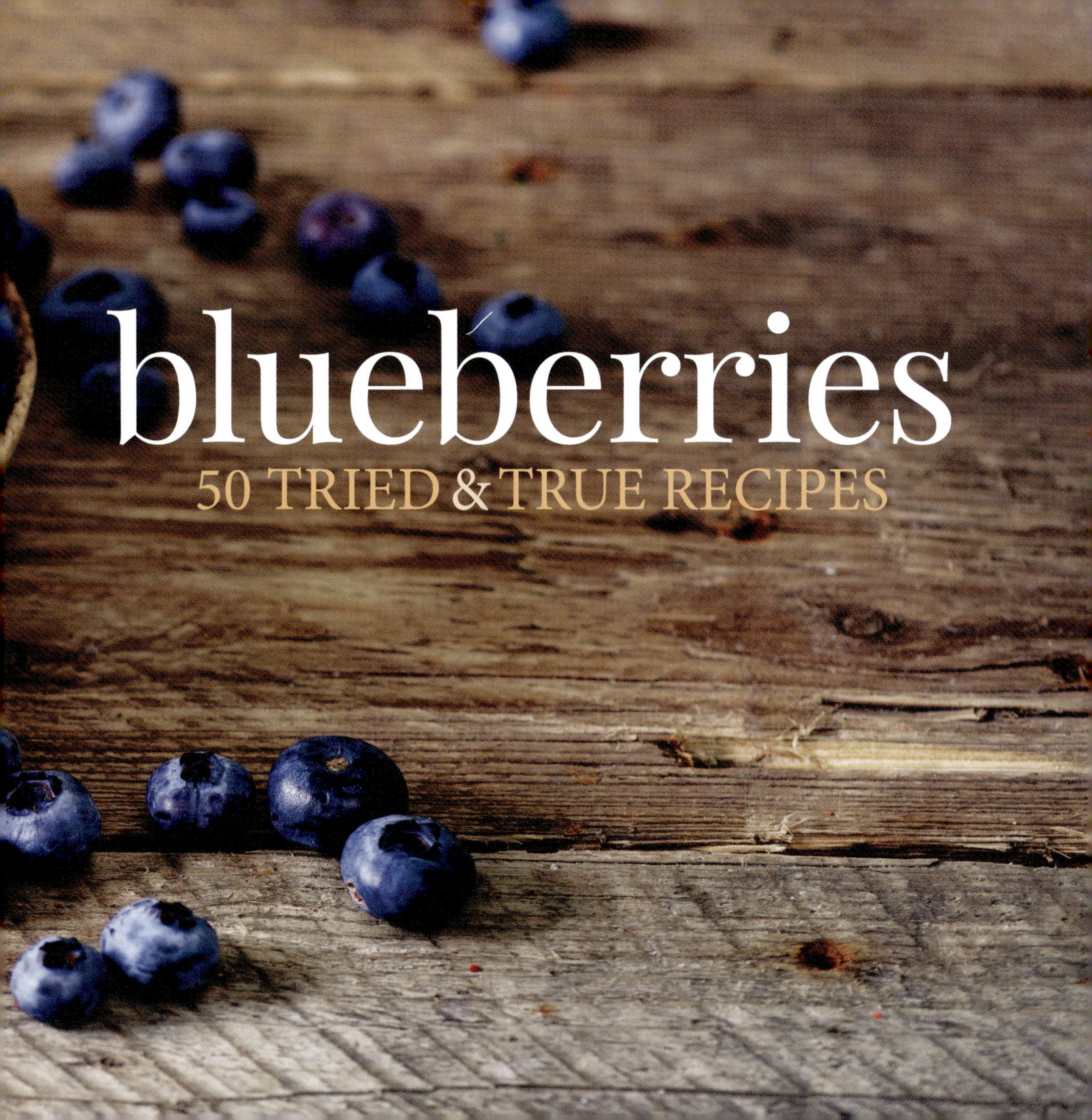

# blueberries

## 50 TRIED & TRUE RECIPES

# Acknowledgments

A bushel of thanks to my dear family: Dit, Emily, and Corinne, who give me the patient support to handle any project, even if it means foraging for dinner when the only hot things on the table are pies, cookies, and turnovers.

To my dear friends Susan Dosier and Jackie Mills—thanks for being my sounding board and text-based support group! You are hundreds of miles away and yet always near and dear. And to my friend Lynne Weems Ryan: Thank you for your delicious contributions and constant generosity.

Thanks to the eager and willing taste-testers of the Ashby Farm Circle Book Club: Paige, Vicki, Sharon (especially for those extra homegrown berries!), Barbara, Connie, Lee, Karen, Jillian, and Rhonda.

To Brett, Emily, and Lora of AdventureKEEN: A big shout-out for helping me create this delightful project that you patiently fine-tuned.

# blueberries

## 50 TRIED & TRUE RECIPES

# Table of Contents

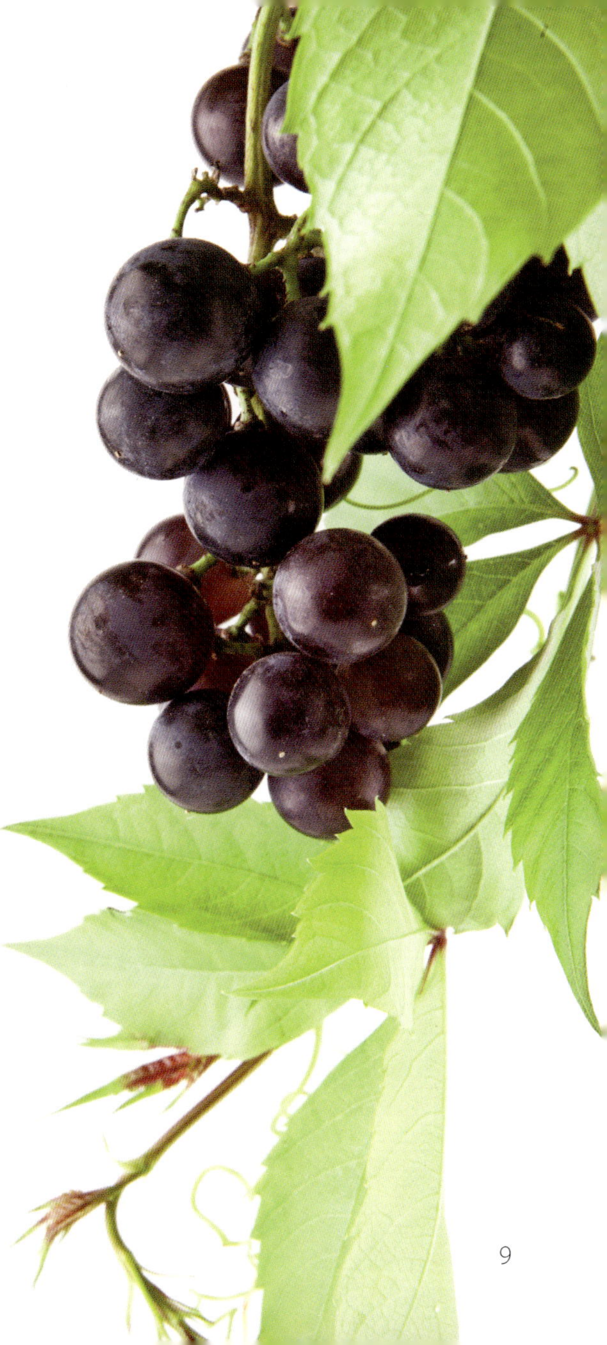

# About Blueberries

### Wild First

It's estimated that wild blueberry plants appeared in Maine more than 10,000 years ago when glaciers receded, and the first scrubby plants grew out of the stressed landscape. Native Americans harvested the berries, followed by European settlers, and yet the crop remained a local bounty. This changed during the Civil War when the berries were harvested, canned, and shipped to feed the Union Army.

### Hybrids Arrive

Skip ahead to 1911, when a farmer's daughter named Elizabeth White teamed up with botanist Frederick Coville to crossbreed and hybridize blueberries that would be suitable as a domesticated, commercial crop.

In 1916, their first commercial crop of highbush blueberries was sold out of Whitesbog, New Jersey.

### Worldwide Success

Commercial blueberry acreage continues to increase, with about half the world's supply coming from the United States. Blueberries are grown commercially in 38 states, but 98 percent of the supply comes from just 10 states: California, Florida, Georgia, Indiana, Michigan, Mississippi, New Jersey, North Carolina, Oregon, and Washington. Maine provides the bulk of the nation's wild blueberry crop.

## Planting and Growing Blueberries

With a small bit of effort, blueberry plants will produce berries for decades. Somewhat easy to grow, blueberry bushes can be used as an ornamental landscape shrub in your yard while supplying sweet white flowers, blue fruit, and then colorful red leaves in the fall.

While blueberries are native to North America, particular varieties are more suited to one region or another. Of the main species of blueberries, there are four main groups: lowbush, Northern highbush, Southern highbush, and rabbiteye.

**Lowbush blueberries** *(Vaccinium angustifolium)* are also called "wild" blueberries, although they are often a carefully managed fruit crop. These native plants are named for their obvious short stature (topping out at less than 2 feet tall), and they can be used as ground cover. They produce small, intensely flavored berries in early summer. Lowbush blueberries thrive in cold climates, in hardiness zones 2–6. Because the plant remains smallish, lowbush blueberries can be grown in containers. Although cold hardy when planted in the ground, potted plants need to be brought under cover during cold spells so the roots don't freeze. Lowbush blueberries spread by seed or underground rhizomes, but limited cultivars are available for purchase.

**Northern highbush** *(Vaccinium corymbosum)* is the most common blueberry variety grown around the world. Most of the berries in grocery stores are from highbush plants. More than half of the highbush blueberries grown are sold fresh in markets; the rest are frozen, canned, or dried. As the name implies, highbush plants can reach heights of 10 to 12 feet. Heat tolerant to zone 7, highbush blueberries require sufficient cold to produce flowers and fruit, so the plants are not suitable for Southern gardens with warm to mild winters. When new varieties are introduced, they're classified by early-, mid-, or late-season.

**Southern highbush** *(Vaccinium corymbosum* hybrid) blueberries were created to bring blueberry production to the South, in climates with hot, humid summers and mild winters. These vigorous plants grow 6 to 8 feet tall and are rated for zones 7–10.

**Rabbiteye** *(Vaccinium ashei* or *Vaccinium virgatum)* blueberries get their name from the way they turn pink before they turn blue, a shade resembling rabbits' eyes. Recommended for hardiness zones 7–10, rabbiteye blueberries are the most heat tolerant and least cold tolerant, making this a popular plant for its native, warm southeastern climate. With a vigorous growth habit, these plants can reach up to 20 feet tall if left unpruned (not recommended, unless you are keen to feeding the birds). Rabbiteyes taste sweeter than Northern highbush varieties, but their skins can get tough when frozen.

**Half-High**, a hybrid cross between Northern highbush and lowbush plants, this type tolerates cold temperatures. These small (3- to 4-feet-high) plants do well in containers.

## Gardening How-Tos

- Purchase at least two different types in the same group (based on climate), and more is better. While most blueberry plants are self-fertile, meaning you don't need another plant to produce berries, cross-pollination with other varieties increases the chance of getting larger berries. If planting several bushes, select some with different fruiting schedules (labeled early-, mid-, or late-season) for a longer harvest time.

- Select a sunny location. While blueberries are somewhat shade tolerant, they will produce more berries when grown in the sun.

- Space plants out based on how large they will become when mature. Plant lowbush 2 feet apart, highbush 5 to 8 feet apart, and rabbiteyes 10 to 15 feet apart.

- Blueberries thrive in acidic, moist, and well-aerated soil. For best results, gardeners should test the pH of the soil and correct to a range of 4.5 to 5.5 with sulfur and other additives. Small, pale leaves and poor growth indicate a pH that is too high (alkaline). Other acid-loving plants include azaleas, rhododendrons, blue hydrangeas, and dogwoods. If those are growing well, then your soil is close to perfect.

- Dig a hole about twice as wide and twice as deep as the roots, filling in with soil amendments such as peat moss, composted leaves, or pine needles. Blueberries are shallow rooted, so make sure not to plant the bush any deeper than it was in the pot. Don't fertilize immediately; wait several weeks before administering a light dose.

- Newly planted blueberries won't need pruning for a few years. Later, prune crowded branches to encourage new branching. (New branches won't produce berries the first year, so don't prune all the plants, or just prune portions of the plants.) Lowbush blueberries should be pruned hard to the ground when production decreases. The traditional method is to burn off fields, which also serves to reduce insects and diseases. Mowing to the ground is often more practical, and loppers can be used if you only have a few plants.

- Birds will steal your harvest. Cover bushes with netting just as the berries begin to ripen, or hang aluminum pie plates from strings to discourage birds and other animals from enjoying a berry buffet.

- Fruit will appear in clusters of 5 to 10 berries, which ripen over a period of weeks. Pick only fully ripe fruit, generally one to three days after the berries turn blue. Waiting to pick results in larger fruit and larger yields.

## Nutrition

- Blueberries are notable for being a healthy "superfood." Early Native Americans made tea from the leaves to treat diarrhea. And flowers and rhizomes were used to treat colic, induce labor, and purify the blood.

- Only 80 calories per cup, blueberries contain practically no fat. They make a good source of fiber at 3.6 grams per serving and an excellent source of manganese (good for weak bones, painful joints, anemia, and premenstrual syndrome). One serving of blueberries equals almost 25 percent of the daily requirement of Vitamin C (important for skin, connective tissue, and promoting healing) and 36 percent of vitamin K (which promotes bone density, cognitive health, and may help reduce blood pressure).

- People are turning to blueberries as a health supplement because the fruit is a rich source of antioxidants. Water-soluble pigments called anthocyanins give blueberries their rich color and offer a plethora of health benefits. In addition to fighting free radicals, anthocyanins are associated with suppressed tumor growth. Other antioxidant activity in blueberries offers cardiovascular protection, anti-diabetic properties, vision improvement, and anti-viral benefits.

## Buying Blueberries

Fresh blueberries are available year-round in markets. Berries grown in the U.S. are available April through October, with peak season running mid-June to mid-August. South America supplies U.S. markets with fresh berries November through March. Fresh "wild" blueberries are a regional specialty with a short season. Most of the crop is sold frozen and is available year-round.

- Look for plump, dark berries with no signs of mold. Look on the bottom of the container for juice stains that may indicate crushed berries. Light or reddish berries indicate underripe fruit.

- Don't be put off by the white sheen on the surface of the berries. The waxy coating, known as "bloom," is a sign of freshness, and it disappears with age and handling. The natural barrier seals in the fruit's moisture and protects against insects and bacteria.

- Taking exact measurements of blueberries can get tricky. More small berries will fit inside a cup measure, which will weigh more than a cup of very large berries. A 6-ounce container holds about ¾ cup, depending on the size of the berries. 1 pint (12-ounce container) yields about 2 cups fresh berries; 1 pound fresh blueberries measures about 3½ cups.

- Blueberries need to be refrigerated or they will quickly mold. Don't wash first; rather, rinse just before using, so dampness doesn't encourage deterioration in the fridge. If you have extra rinsed berries, dry them thoroughly on paper towels before refrigerating. Fresh blueberries will keep about 1 week.

## Freezing Blueberries

Long-term storage couldn't be simpler—just place berries in a zip-top plastic bag or airtight container and pop it in the freezer! Even better, lay out washed berries on paper towels to dry. Place berries in a single layer on a sheet pan and place in freezer. Once berries are frozen, transfer to a zip-top plastic freezer bag; store up to 10 months. There is nothing to peel or chop. Frozen blueberries can almost always be substituted for fresh. It's a good thing, too, because fresh berries can be expensive and the season is short in some areas. As easy as it is, here are some tips.

- Keep 'em dry: If you've washed the berries, lay them out on a towel to air-dry. If damp, the berries will stick together in clumps when freezing and can introduce ice crystals, then water, into recipes.

- Pre-measure: Instead of a bag of unknown quantity, measure berries by the cupful and keep each bag the same amount. Label with the date and quantity.

- Frozen berries in commercial bags should feel loose and not clumped together. Blueberries that have been individually quick-frozen are as easy to measure as fresh ones. Measure berries soon after removing from the freezer; once thawed, blueberries may collapse and sink into measuring cups.

## Cooking with Fresh or Frozen Blueberries

- Frozen blueberries can be substituted for fresh in most recipes, with a few exceptions. Cooked sauces and fillings that are later thickened with cornstarch, flour, or tapioca can be made with frozen blueberries, which thaw, then reduce, when heated.
- Pies baked with raw berries sometimes don't slice well or thicken if the berries were previously frozen. It depends on the berries, water content when frozen, any residual ice crystals, etc. If you must use frozen berries in uncooked pies and tarts, measure and then thaw on paper towels to reduce extra weeping.
- Cakes, muffins, pancakes, etc., that include a good deal of flour or dry ingredients can handle the extra moisture a frozen berry might introduce into the batter.
- Frozen berries reduce the core temperature of foods, so you may need to increase baking time 5, 10, or more minutes, until food is done in the center.
- Neutral pH keeps blueberries at a purplish tint. Adding acid (such as lemon or lime juice) to recipes reduces the pH, changing the berries to a pinkish hue. On the other hand, a high pH (a result of adding baking soda) casts a green tint to foods. Don't be alarmed if smoothies or recipes made with dairy have a slightly grayish cast. Add a bit of lemon juice to brighten them up.
- Gently fold blueberries into recipes last, to avoid streaking the batter.
- Some recipes recommend tossing the blueberries in a bit of flour to help suspend them in the middle; otherwise, the berries might sink to the bottom before the item finishes baking. Or spread half the batter in the pan, top with half the berries, and then spread the remaining batter and berries so they are distributed evenly.

## So, What About Dried?

Markets sell both wild and highbush dried blueberries in small, shelf-stable packages. Look for them at your grocery store with the raisins and dried cranberries. Many brands include unnecessary sugar or high-fructose corn syrup. Unsweetened ones can be found at stores that stock a good variety of natural and organic products. Unsweetened berries are sweet enough on their own and add a potent kick of flavor in a small bite. Except for jams and pies, you can substitute dried blueberries at about ½ to ¾ the amount of fresh (because they are about half the size of fresh or frozen).

## What's the Difference?

Before you start baking, it's handy to know the names of a few popular blueberry dishes.

Crumble: a baked dish of fruit filling topped with an oat-based sweet streusel topping, sometimes made with graham crackers, flour, and nuts; held together with butter

Crisp: same as a crumble but without the oats; sometimes made with graham crackers or a simple flour-sugar-butter streusel

Brown Betty: same non-oat topping as a crisp, but the sweetened fruit mixture is layered in between the streusel mixture, with fruit in the middle

Cobbler: a deep-dish fruit dessert with individual biscuit dough dropped—or cake batter spread—on top before it's baked in an oven

Slump or Grunt: same biscuit or dumpling topping as a cobbler but cooked in a covered skillet on a cooktop

Buckle: a single layer cake with fruit stirred into the batter and baked in the oven. Often has a streusel topping or cake batter under the fruit. As it bakes, the fruit sinks while the cake batter rises around the fruit.

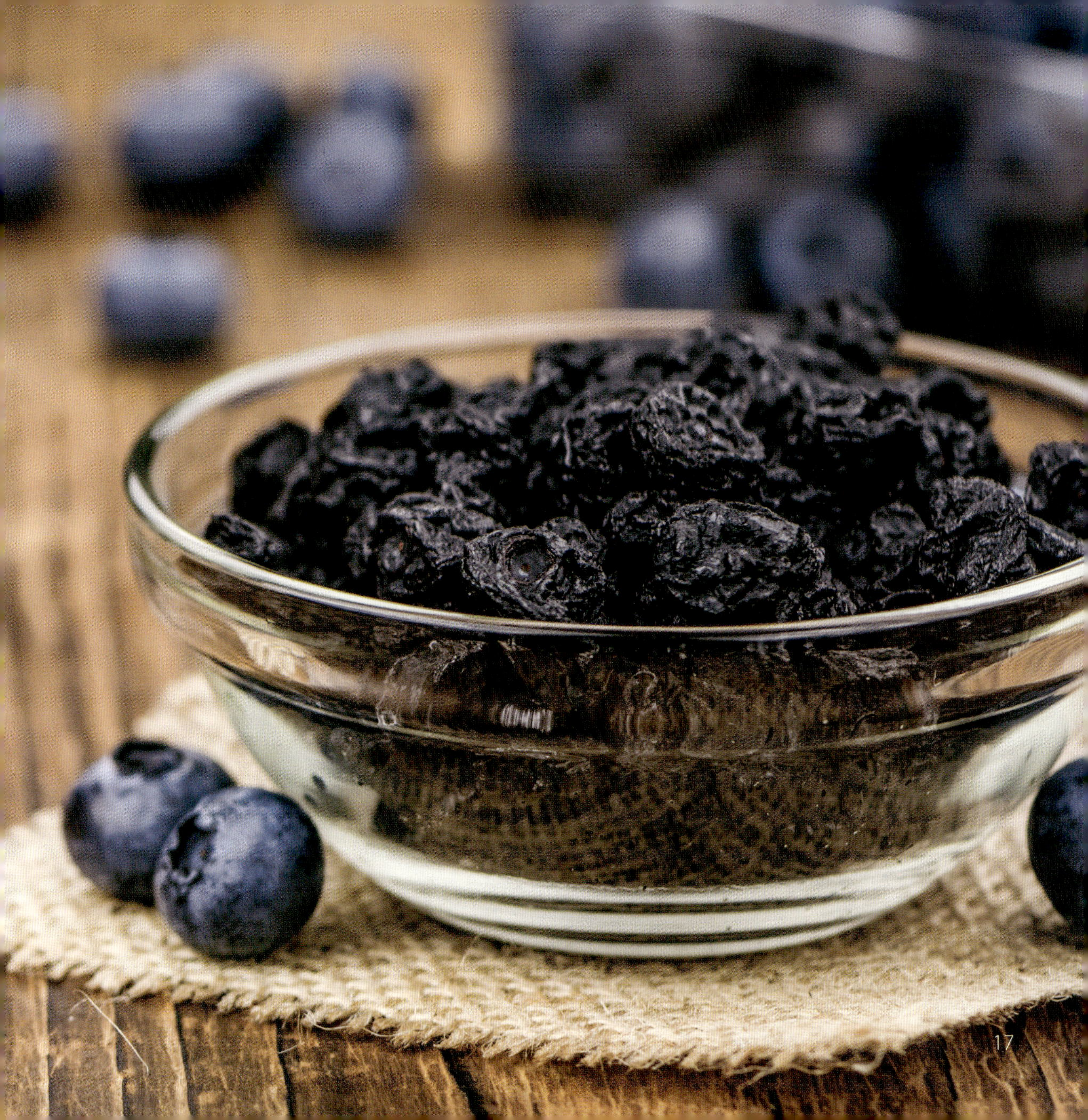

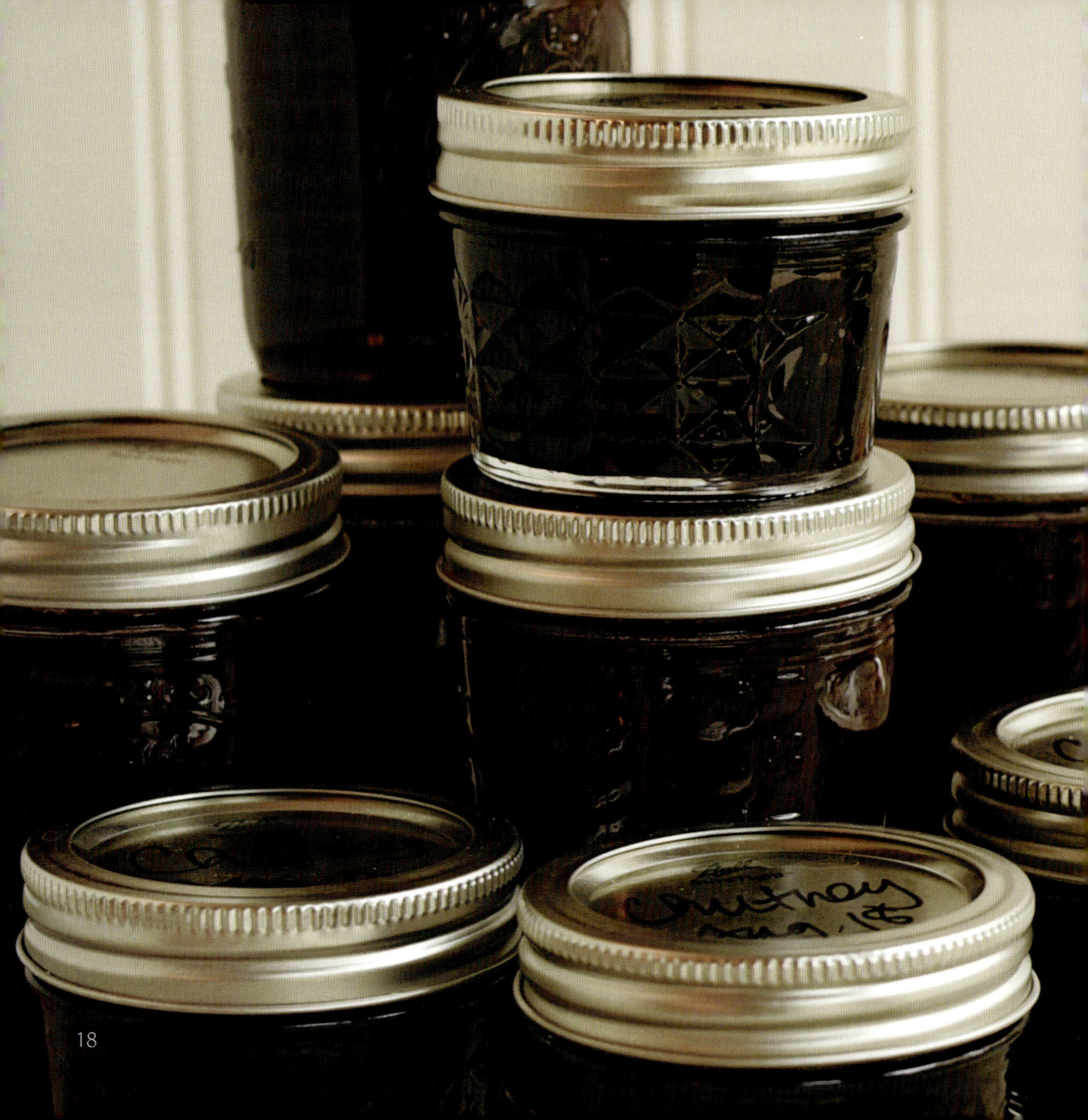
Chutney
Aug '15

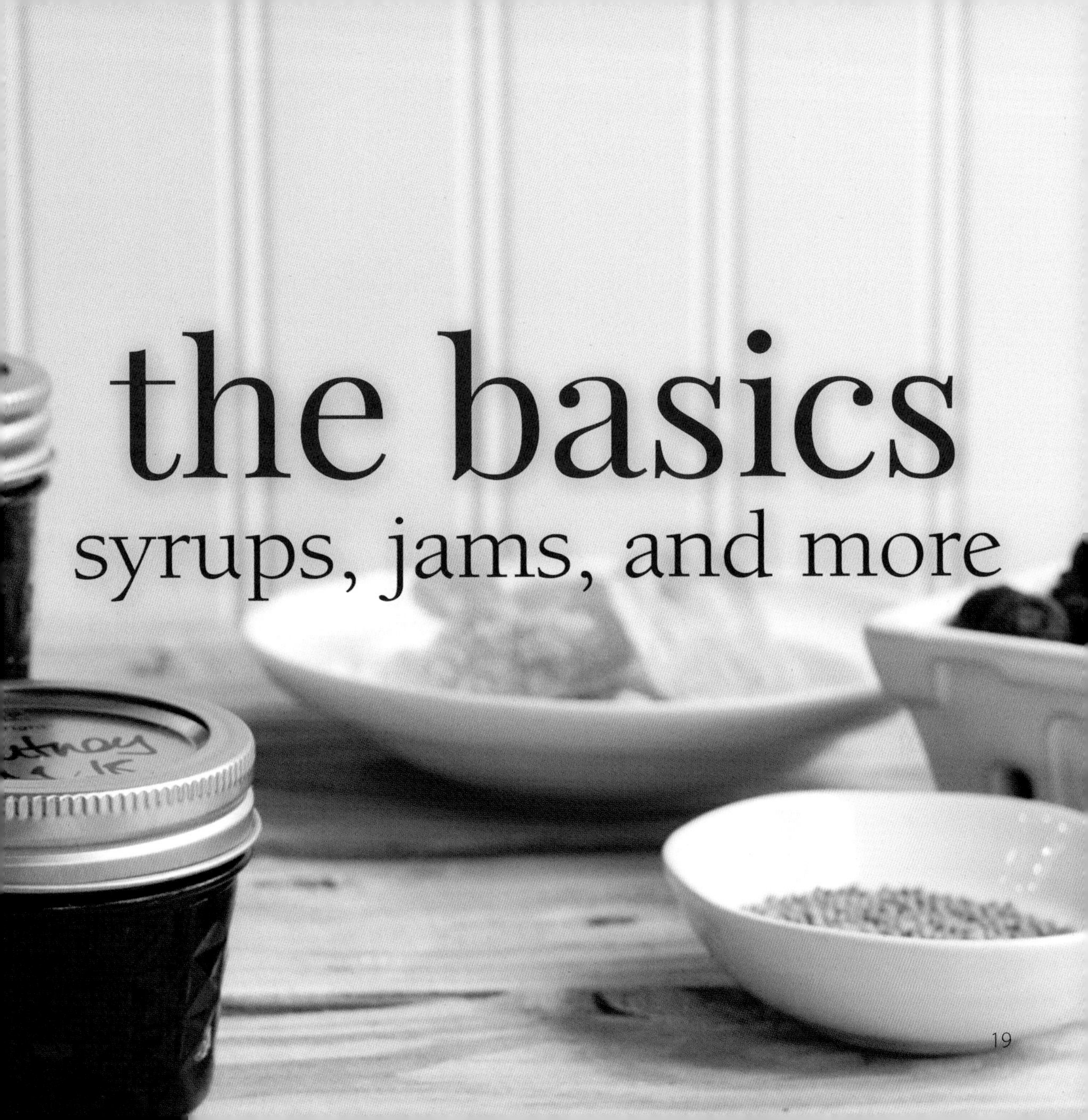

# the basics

## syrups, jams, and more

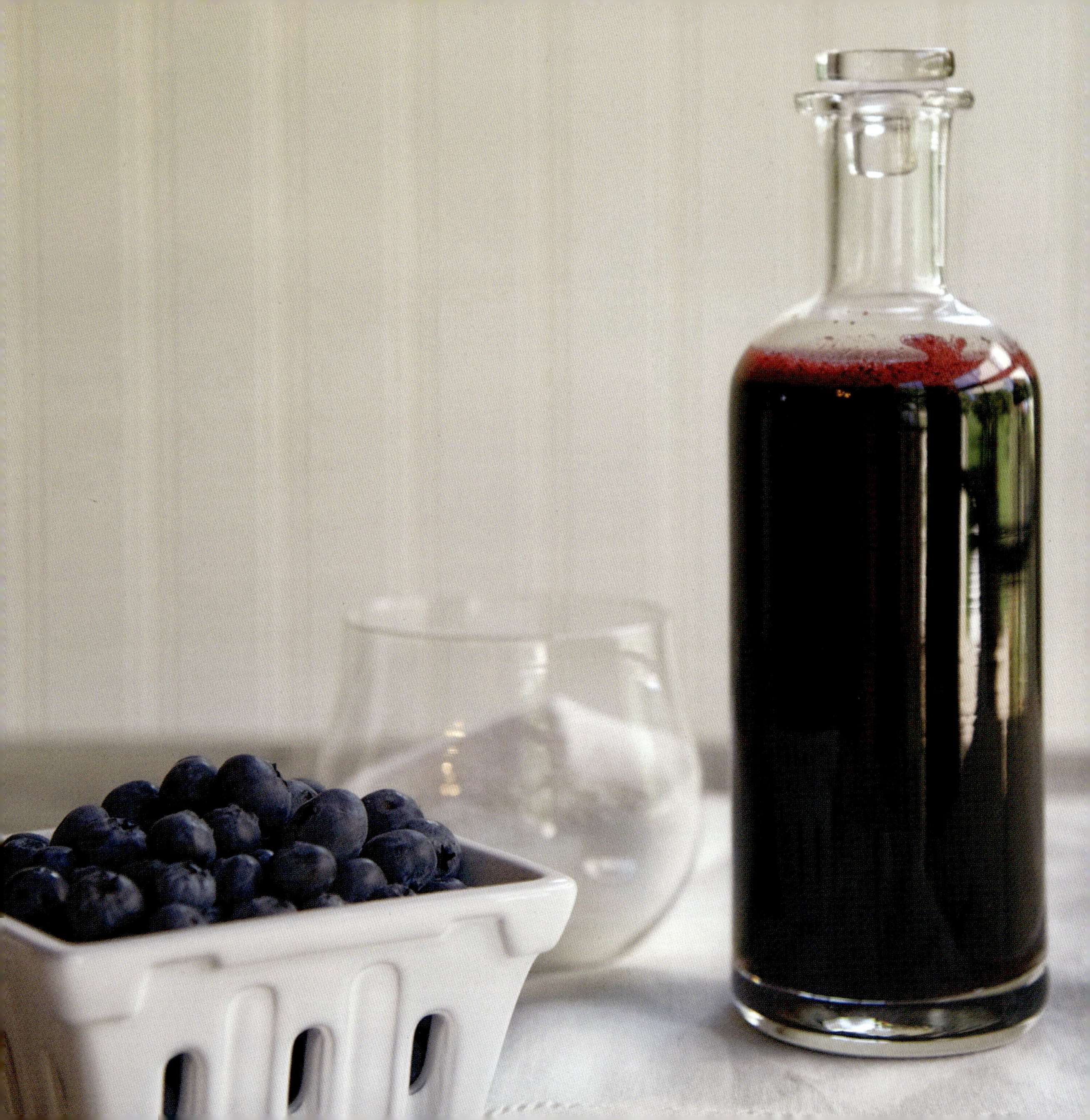

# Simple Blueberry Syrup

Use this versatile syrup as a sauce for cakes or ice cream. Try it as a fruity sweetener in sparkling water spritzers or in vodka or gin cocktails for a light summer drink.

makes 2 cups

INGREDIENTS

1 cup granulated sugar
1 cup water
3 cups fresh or frozen blueberries

Combine sugar, water, and blueberries in a nonreactive saucepan over medium-low heat. Cook 3 minutes, stirring frequently, until sugar dissolves. Increase heat to medium. Simmer 10 minutes, crushing blueberries with back of spoon as they soften. Cool slightly.

Strain mixture through a fine wire-mesh sieve set over a bowl, pressing mixture with spoon or spatula. Discard pulp. Transfer to a glass bottle or jar. Refrigerate until ready to serve.

*Vanilla Blueberry Syrup:* Cut **1 vanilla bean** in half lengthwise, using the back of a paring knife to scrape out the seeds; stir seeds into blueberry mixture in saucepan before cooking. Place vanilla bean into blueberry mixture (cut in half if too large to fit); let steep 15 minutes, and proceed as directed. Discard vanilla bean with pulp before transferring to a glass bottle or jar.

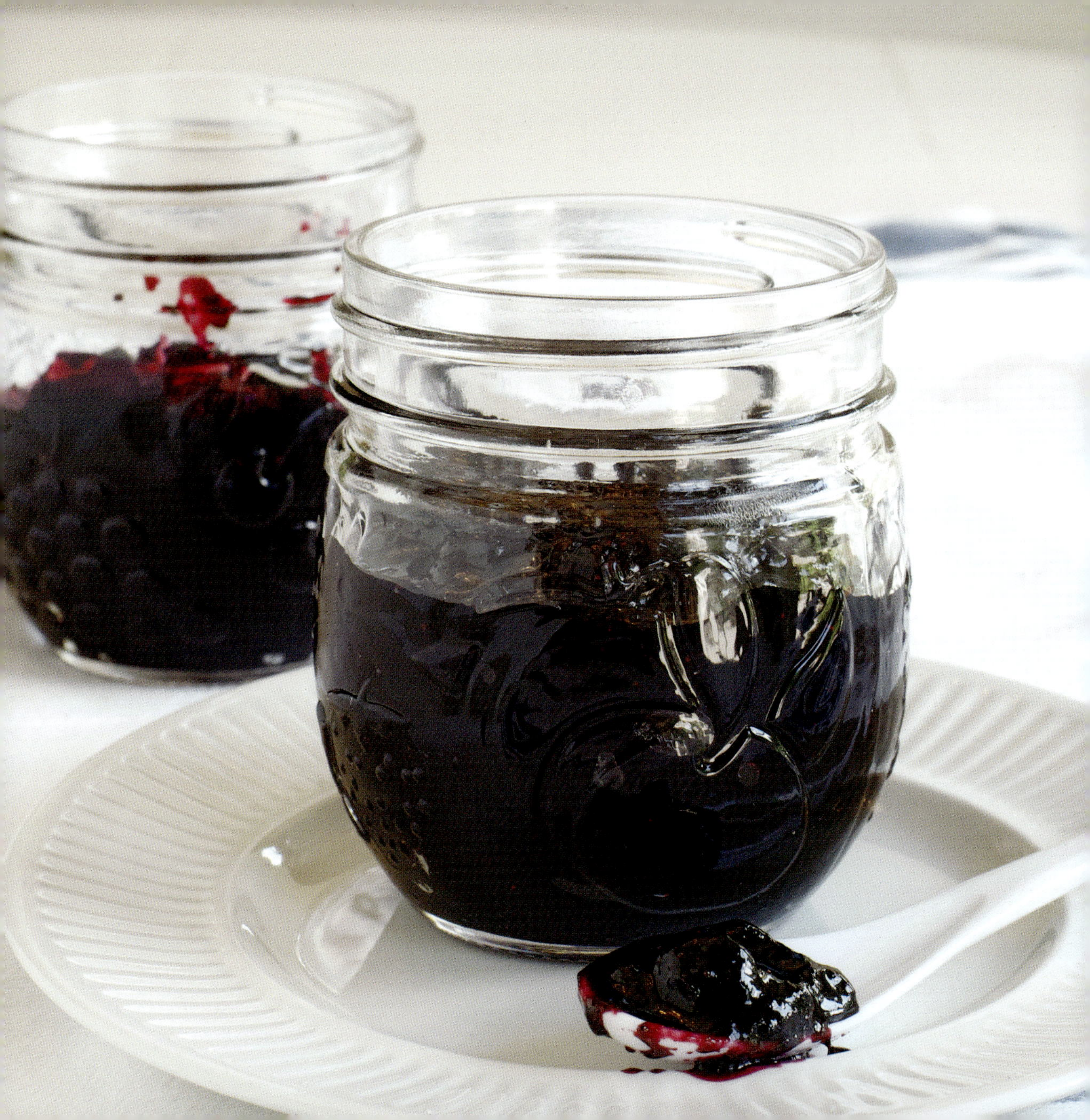

# Simplest Blueberry Jam

This easy recipe uses time and heat to thicken the berries into a rich jammy texture. The mixture reduces to a small amount that's not really worth processing in canning jars. Just keep in the fridge and enjoy while it lasts.

makes 1¼ cups

INGREDIENTS

4 cups fresh or frozen blueberries
1 cup granulated sugar
1 tablespoon fresh lemon juice
⅛ teaspoon kosher salt

Combine blueberries, sugar, lemon juice, and salt in a large saucepan. Cook over medium heat for 25 to 30 minutes, crushing berries with back of spoon to release juice. Mixture will be thick and chunky.

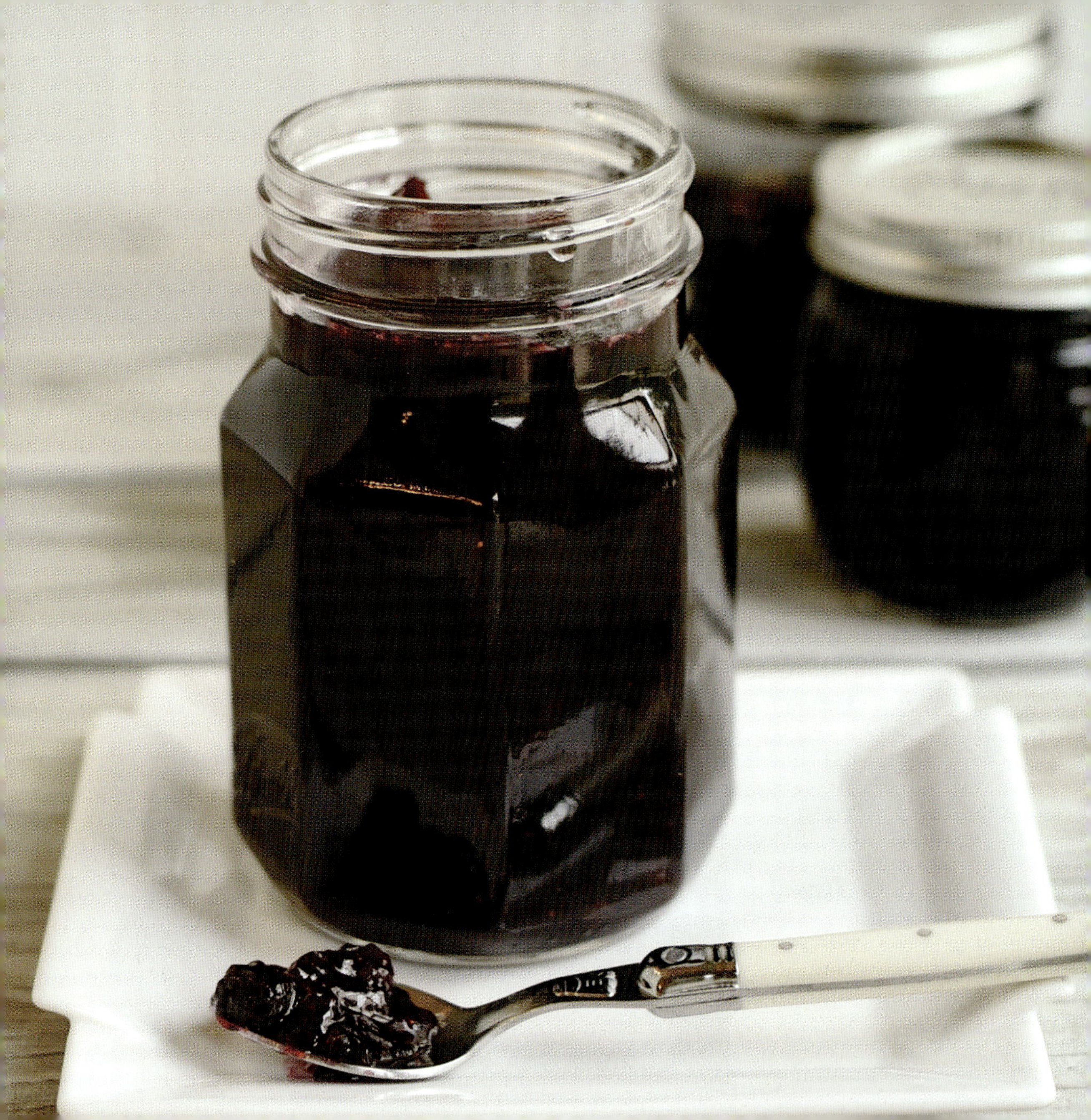

# Blueberry-Orange Marmalade

Peeling and boiling citrus rind is time-consuming. This hybrid recipe substitutes blueberries for some of the citrus rind and pulp. Baking soda softens rinds without lengthy boiling. Blueberries don't contain as much pectin as citrus fruit; because I prefer a firmer texture, I add pectin to make sure the mixture gels. Buy organic citrus because the peel is eaten.

makes 6 cups

INGREDIENTS

2 organic seedless oranges
1 organic lemon
1½ cups water
⅛ teaspoon baking soda
4 cups fresh or frozen blueberries, divided
1 tablespoon butter
5 cups granulated sugar
1 (6-ounce) package liquid fruit pectin

Peel oranges and lemon with a knife or vegetable peeler; cut away most of white pith and discard. Slice rind into thin pieces. Bring 1½ cups water to a boil in a large stock pot; stir in rind and baking soda. Simmer over low heat for 15 minutes, stirring occasionally. (Don't allow water to boil away. If mixture gets dry, add a cup of water.)

Remove excess pith from peeled citrus; cut fruit into large chunks, removing any seeds. Place in a food processor and process until finely chopped. Add 2 cups blueberries; process until coarsely chopped.

Add orange-blueberry mixture, remaining 2 cups blueberries, and butter to rind in pot; simmer over medium-low heat for 20 minutes. Add sugar and cook, stirring constantly, until sugar dissolves. Increase heat to medium-high and bring to a boil. Boil without stirring for 5 minutes.

Remove from heat and stir in pectin. Skim away any foam, and let mixture settle 5 minutes to allow fruit to sink from the top (do not allow to cool completely). Pour hot mixture into sterilized jars, filling to ¼ inch from the top. Remove any air bubbles and wipe rims. Cover with metal lids and screw on bands. Process in boiling water bath for 10 minutes.

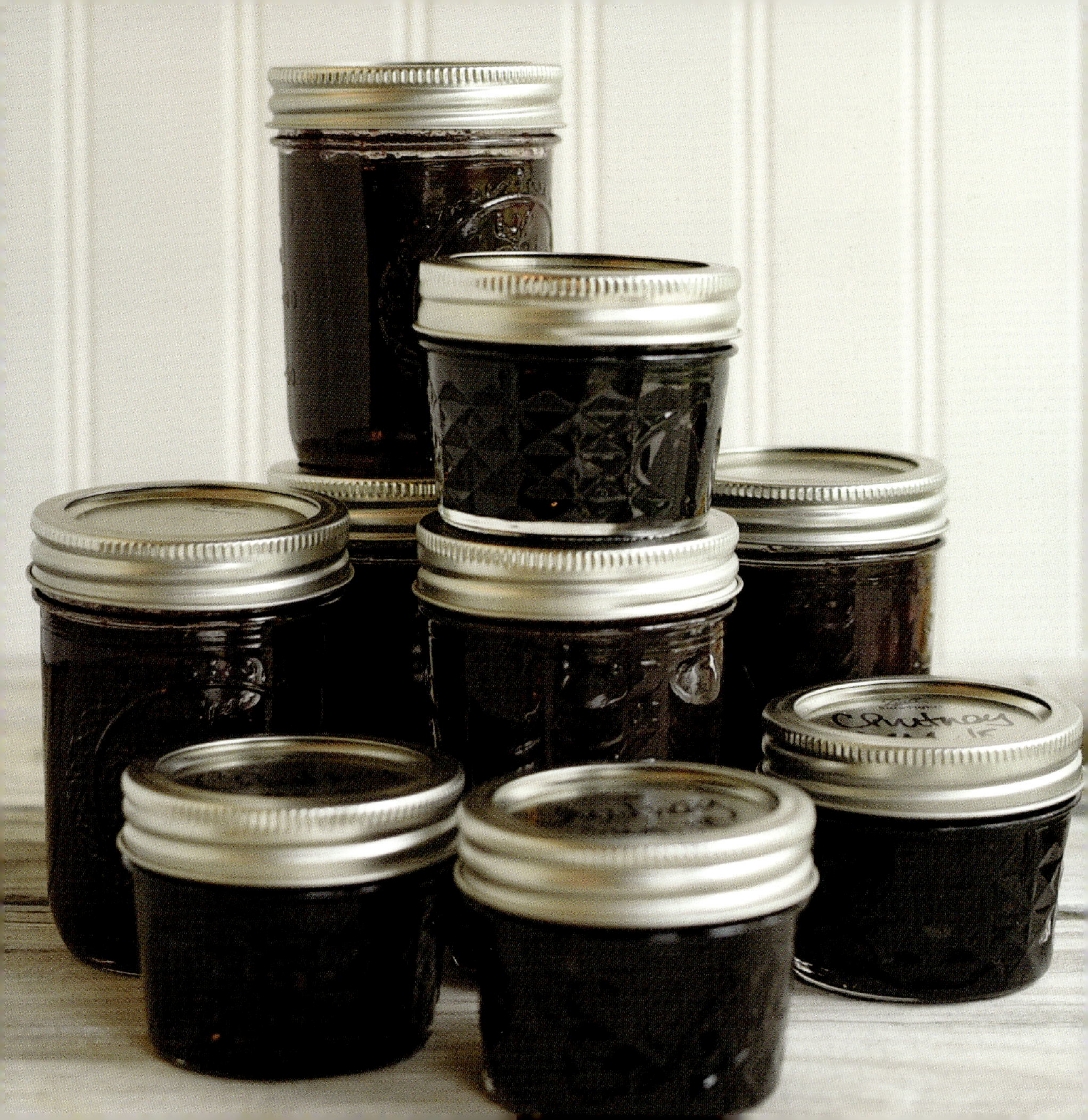
Chutney

# Blueberry-Ginger Chutney

This spicy condiment isn't "hot," even though it contains red pepper flakes. The rich and tangy flavor pairs well with meats. For an easy but special appetizer, pour this chutney over a block of cream cheese, and serve with crackers or gingersnaps.

makes 3 cups

INGREDIENTS

1 cup apple cider vinegar
1 cup granulated sugar
1 cup firmly packed light brown sugar
1 tart apple, peeled and chopped
½ cup minced crystallized ginger
1 tablespoon mustard seeds
2 teaspoons lemon zest
½ teaspoon ground cinnamon
½ teaspoon salt
¼ teaspoon crushed red pepper flakes
¼ teaspoon ground cloves
6 cups fresh or frozen blueberries

Combine vinegar, sugars, apple, ginger, mustard seeds, zest, cinnamon, salt, red pepper flakes, and cloves in a large pot over medium heat. Cook, stirring frequently, until sugar dissolves.

Increase heat and bring mixture to a boil. Reduce heat to medium-low and simmer 30 minutes until mixture is reduced and slightly syrupy.

Stir in blueberries. Bring to a simmer and cook 45 minutes, stirring periodically. Transfer to canning jars and store in refrigerator. For shelf-stable storage or gifts, process for 10 minutes in boiling water.

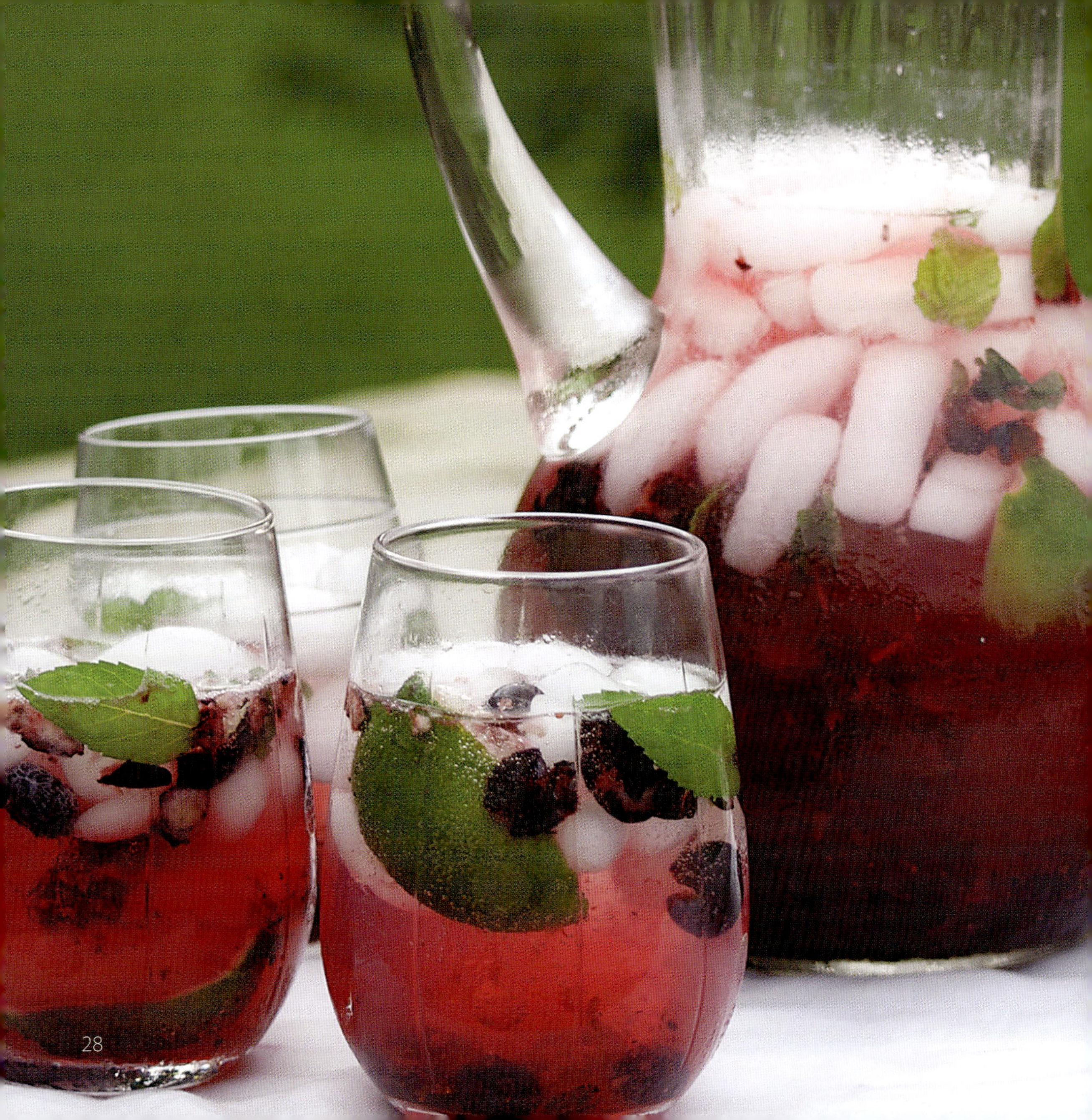

# beverages

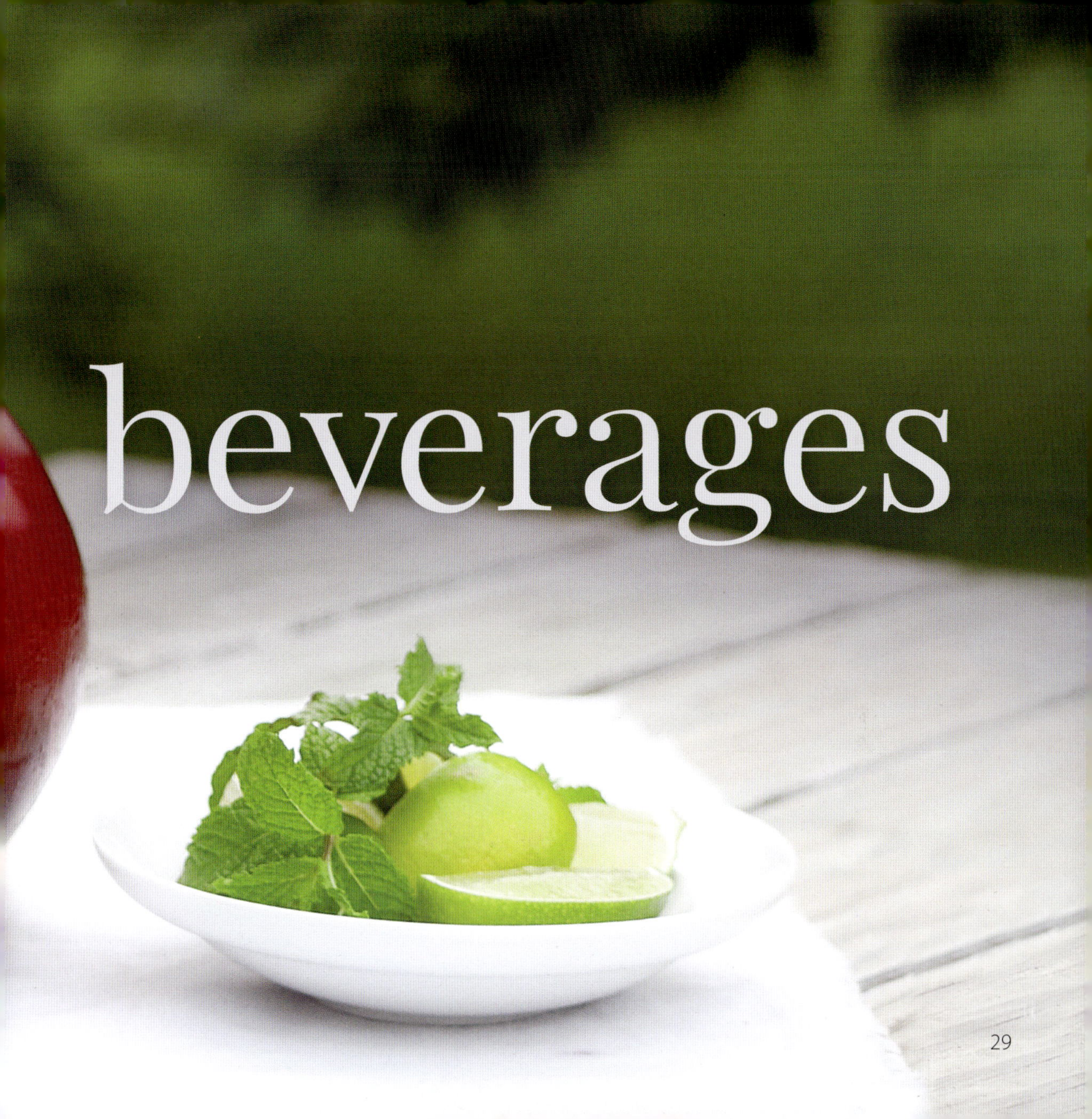

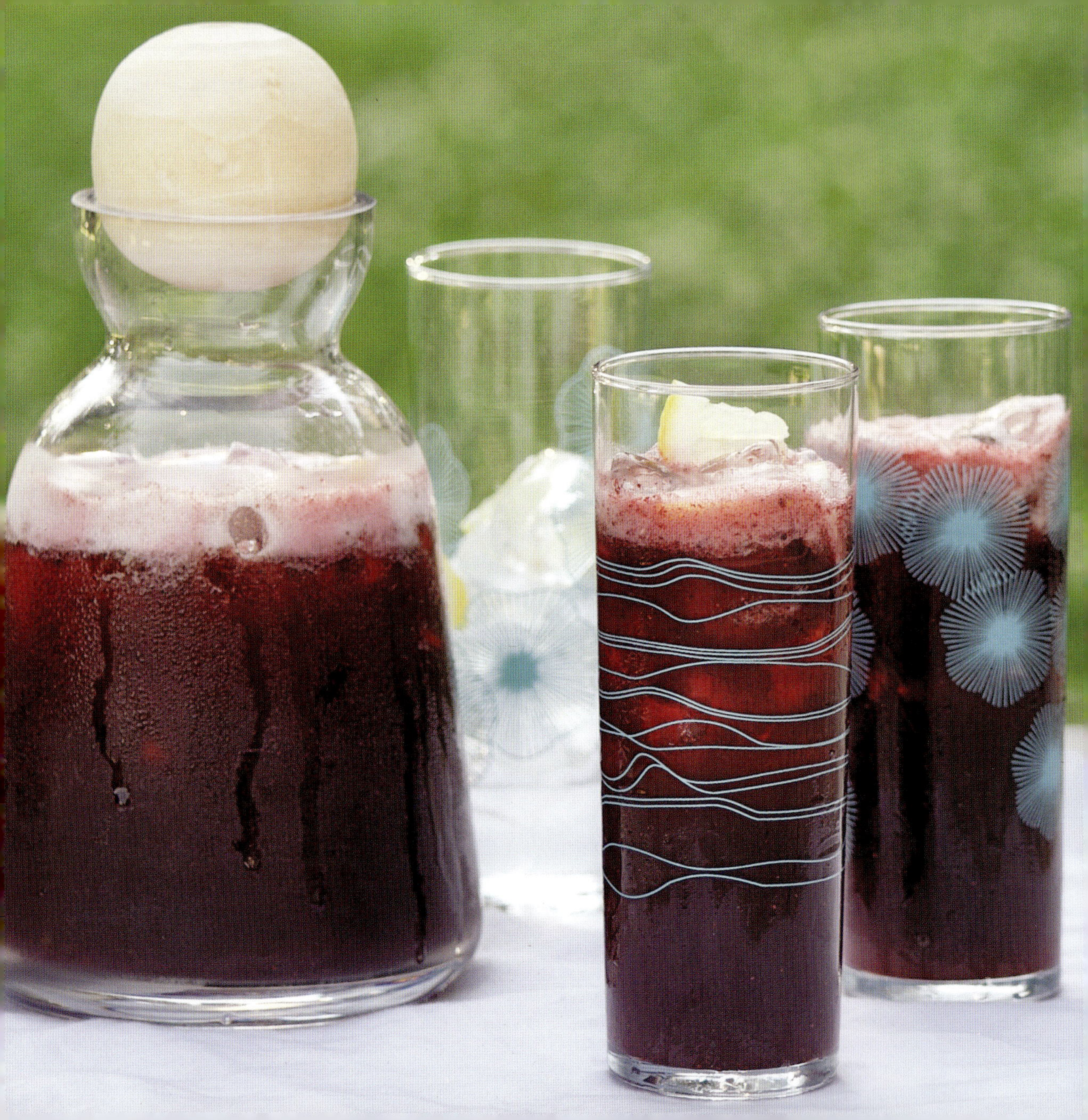

# Blueberry Lemonade

Lemonade makes the perfect summer beverage, and seasonal blueberries add extra flavor to this refreshing drink.

makes 4 cups

INGREDIENTS

2 cups fresh or frozen blueberries
1 cup granulated sugar
1 cup fresh lemon juice
2 cups water

Combine blueberries, sugar, and juice in a blender. Process until smooth. Strain mixture through a fine wire-mesh sieve set over a bowl, pressing mixture with a spoon or spatula. Discard pulp. Add water; serve over ice.

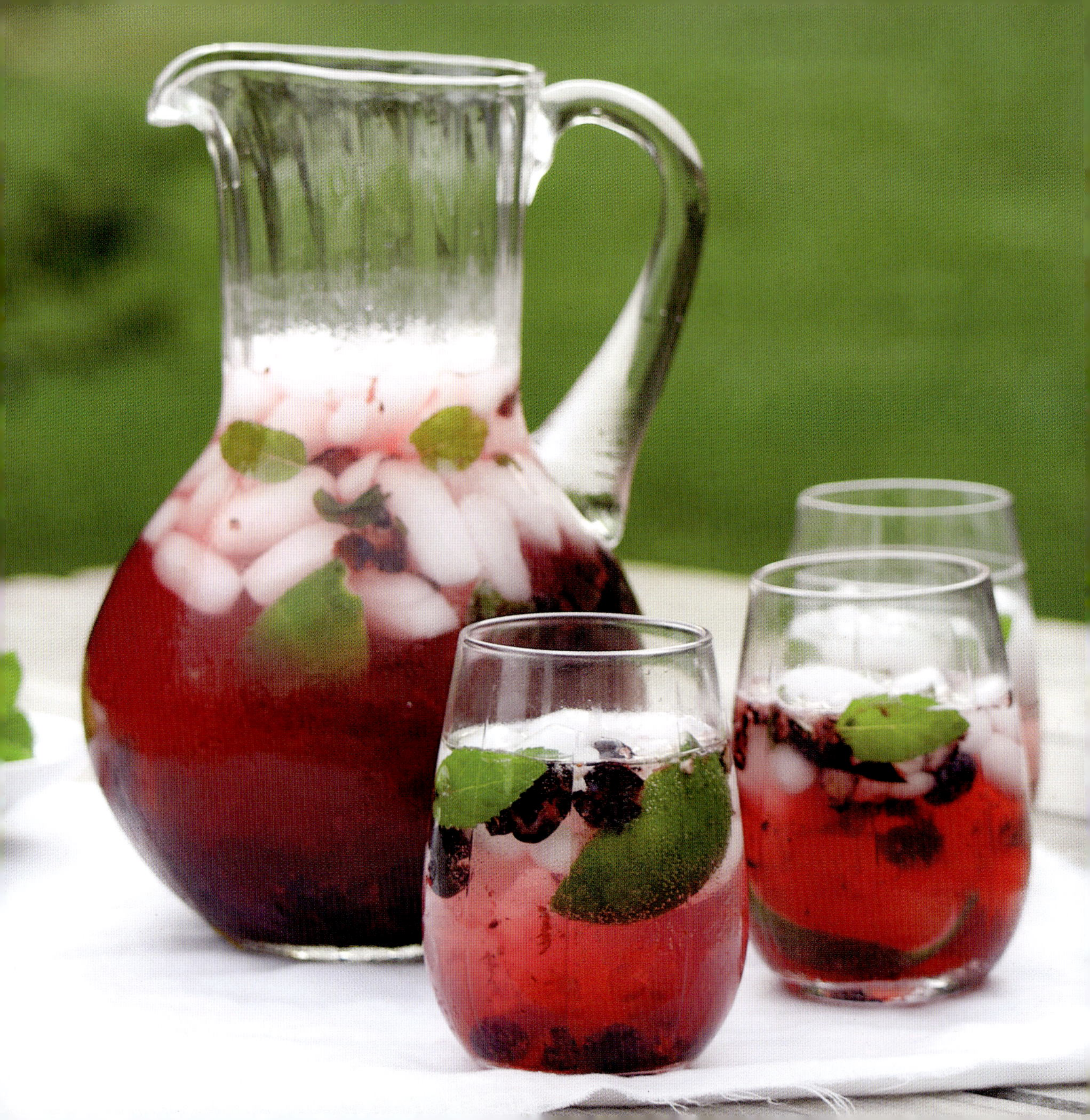

# Blueberry Pitcher Mojito

Thaw frozen berries for a few minutes for easier muddling.

makes 6 servings

**INGREDIENTS**

½ cup loosely packed fresh mint leaves, torn
1 cup fresh or frozen blueberries
½ cup fresh lime juice
⅓ cup Simple Syrup (recipe at right)
1½ cups white rum
4 to 6 cups club soda or sparkling water

**GARNISHES**

fresh blueberries, fresh mint, lime wedges

Combine mint and blueberries in a bowl or pitcher. Press vigorously with a muddler or wooden spoon until berries are crushed and mint is bruised. Stir in lime juice, Simple Syrup, and rum. Cover and chill until ready to serve.

To serve, gently stir club soda into pitcher. Serve in ice-filled glasses. Garnish, if desired.

*Simple Syrup:* Combine **2 cups water** and **2 cups sugar** in a small saucepan over medium heat. Cook, stirring occasionally, until sugar is dissolved. Cool; transfer to a jar and refrigerate until ready to use.

*Blueberry Mojito for Two:* Divide **¼ cup torn fresh mint leaves**, **¼ cup blueberries**, and **4 lime wedges** in 2 glasses. Crush vigorously with a muddler or wooden spoon until mint is bruised and berries and lime are juiced. Fill glasses with ice and stir in **¼ cup Simple Syrup** and **¼ cup rum**. Stir in **sparkling water**. Garnish, if desired.

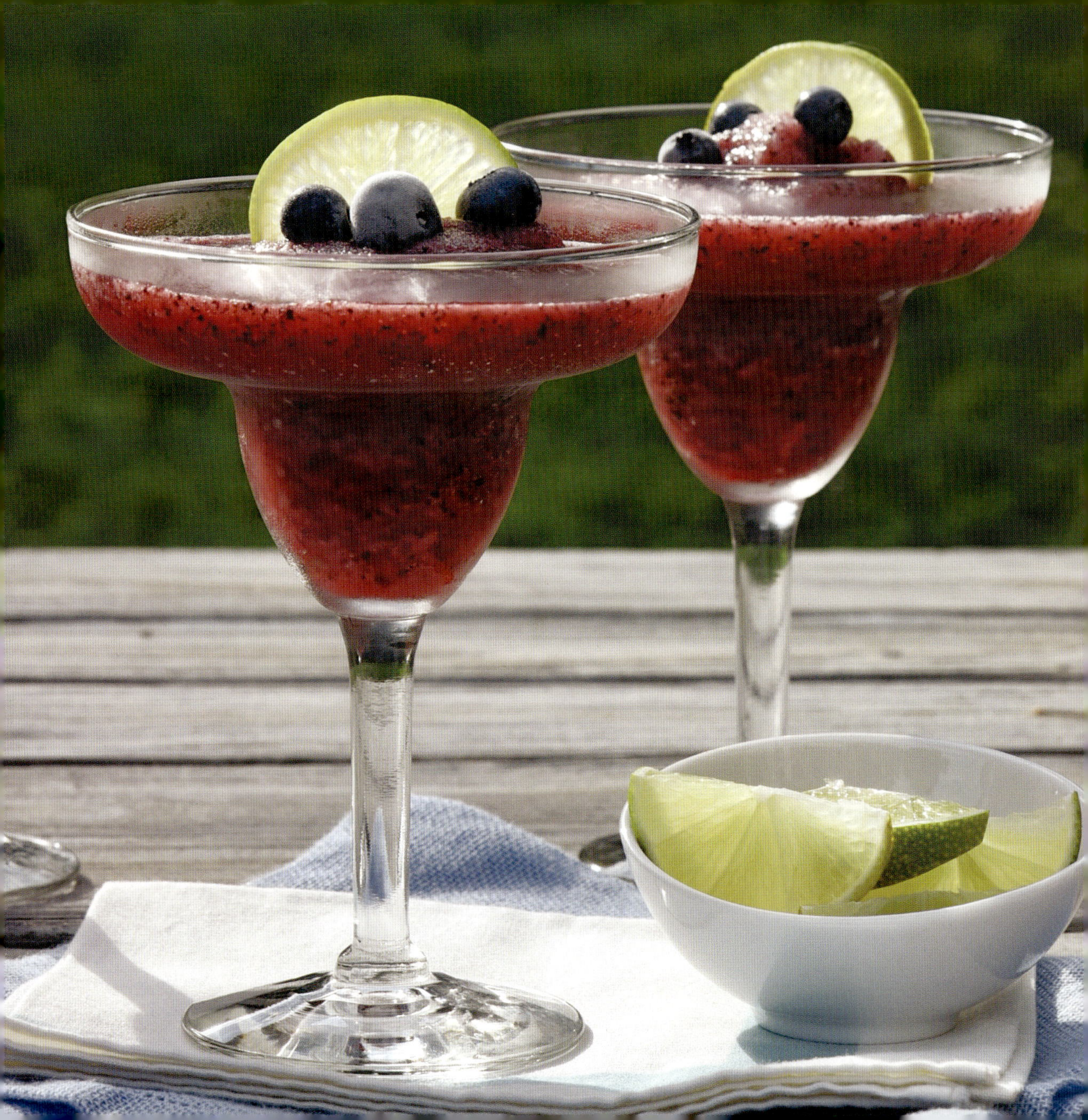

# Blended Blueberry Margarita

There's a lot of wiggle room in this recipe. If berries are tart, taste and then add a bit more Simple Syrup. And adjust the amount of tequila with less if you want a mild cocktail, more for a potent kick. For a fun and slushy blueberry limeade, omit liqueurs and add a bit of water to make it a drinkable texture.

makes about 6 cups

INGREDIENTS

2 cups fresh or frozen blueberries
1¼ cups Simple Syrup (recipe at right)
¾ cup fresh lime juice
¾ cup tequila
¼ cup triple sec or other orange liqueur
6 cups ice or crushed ice

GARNISHES

lime wedges, fresh blueberries

Combine blueberries, Simple Syrup, and lime juice in a heavy-duty blender. Process until berries are finely chopped and mixture is almost smooth.

Stir in tequila, triple sec, and ice. Process until ice is finely crushed and mixture is slushy. Pour into glasses; garnish, if desired.

*Simple Syrup:* Combine **2 cups sugar** and **2 cups water** in a saucepan over medium heat. Cook a few minutes, stirring occasionally, until sugar dissolves. Cool; transfer to a jar and refrigerate until ready to use.

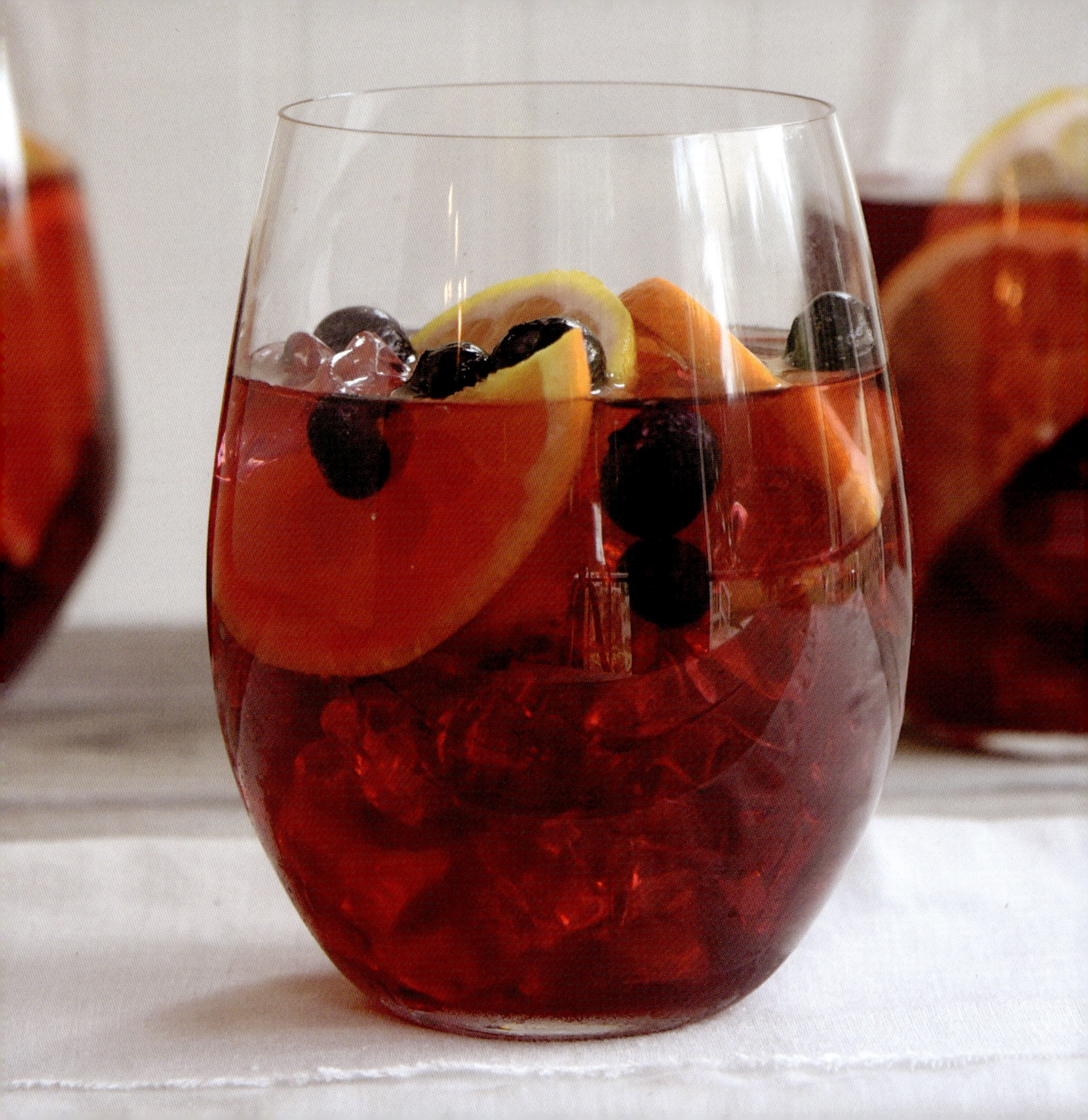

# Blueberry-Lemonade Sangria

This fun, summery adult beverage takes a shortcut with packaged lemonade for a really zingy drink. For a more mellow flavor, you can stir in 2 cups refrigerated prepared lemonade.

makes 8 cups

**INGREDIENTS**

1½ cups fresh or frozen blueberries
1 (12-ounce) container frozen lemonade concentrate
1 cup prepared blueberry juice or blueberry-cranberry juice, chilled
1 (1.5-liter) bottle Sauvignon Blanc or other dry white wine, chilled
½ cup triple sec or other orange liqueur (optional)
1 seedless orange, thinly sliced
1 lemon, thinly sliced

Combine blueberries and lemonade concentrate in a large pitcher. Press with a spoon to crush some of the berries. Stir in juice, wine, and, if desired, liqueur. Stir in orange and lemon slices. Cover and chill 2 to 8 hours before serving.

To serve, pour over ice in glasses.

*Note:* For a light, sparkling beverage, fill glasses ¾ full with sangria, and top with sparkling water.

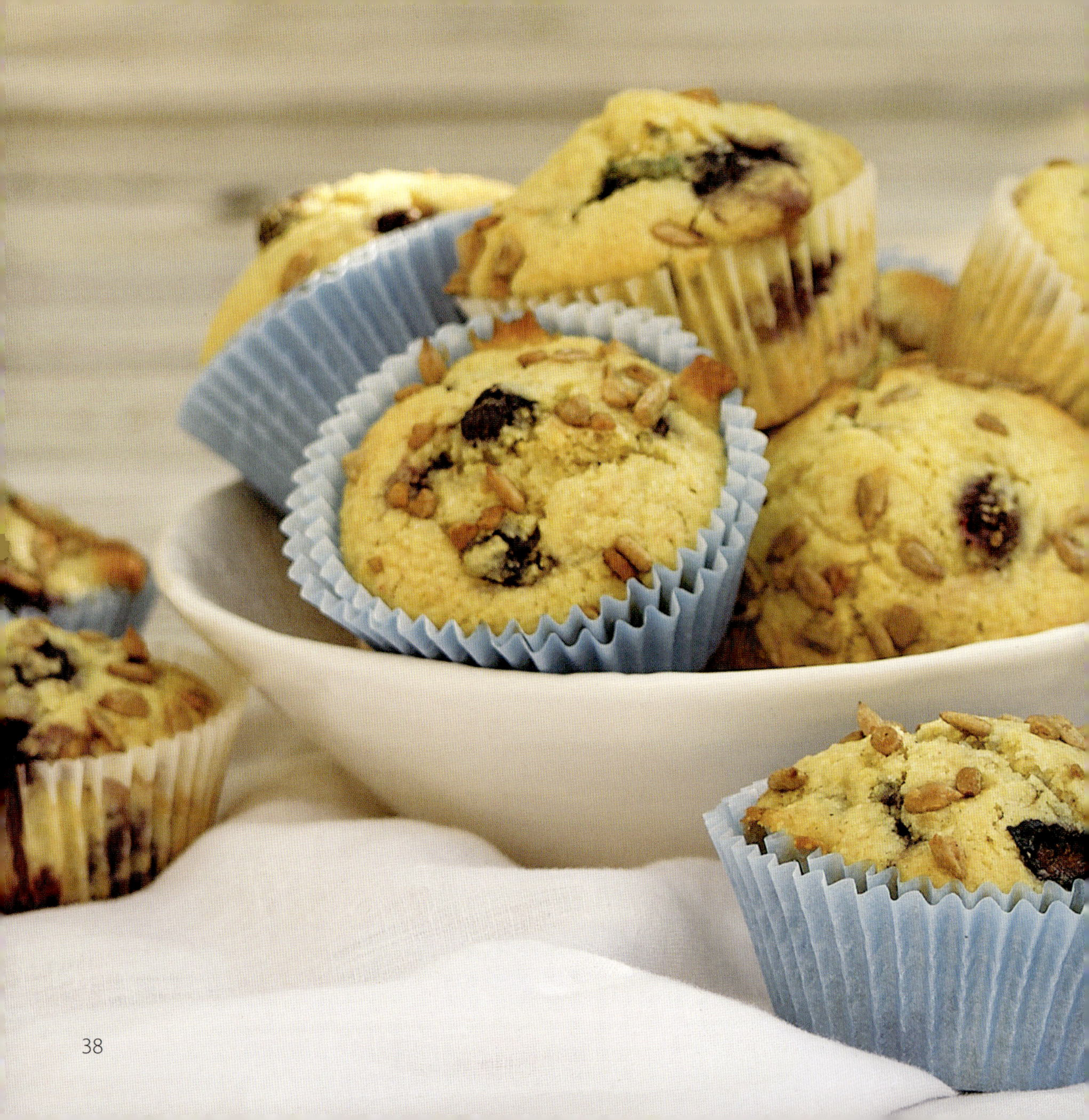

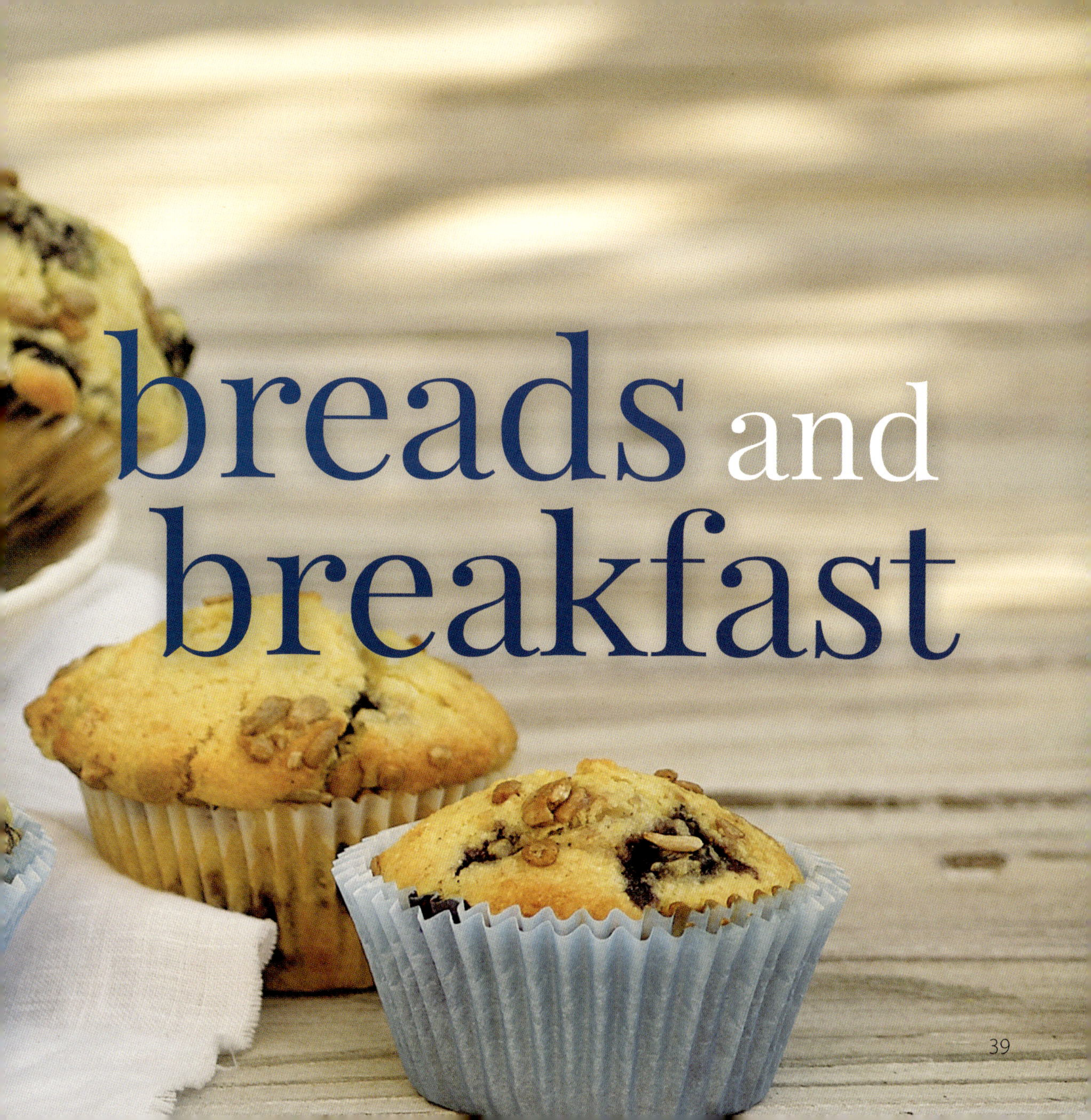

# breads and breakfast

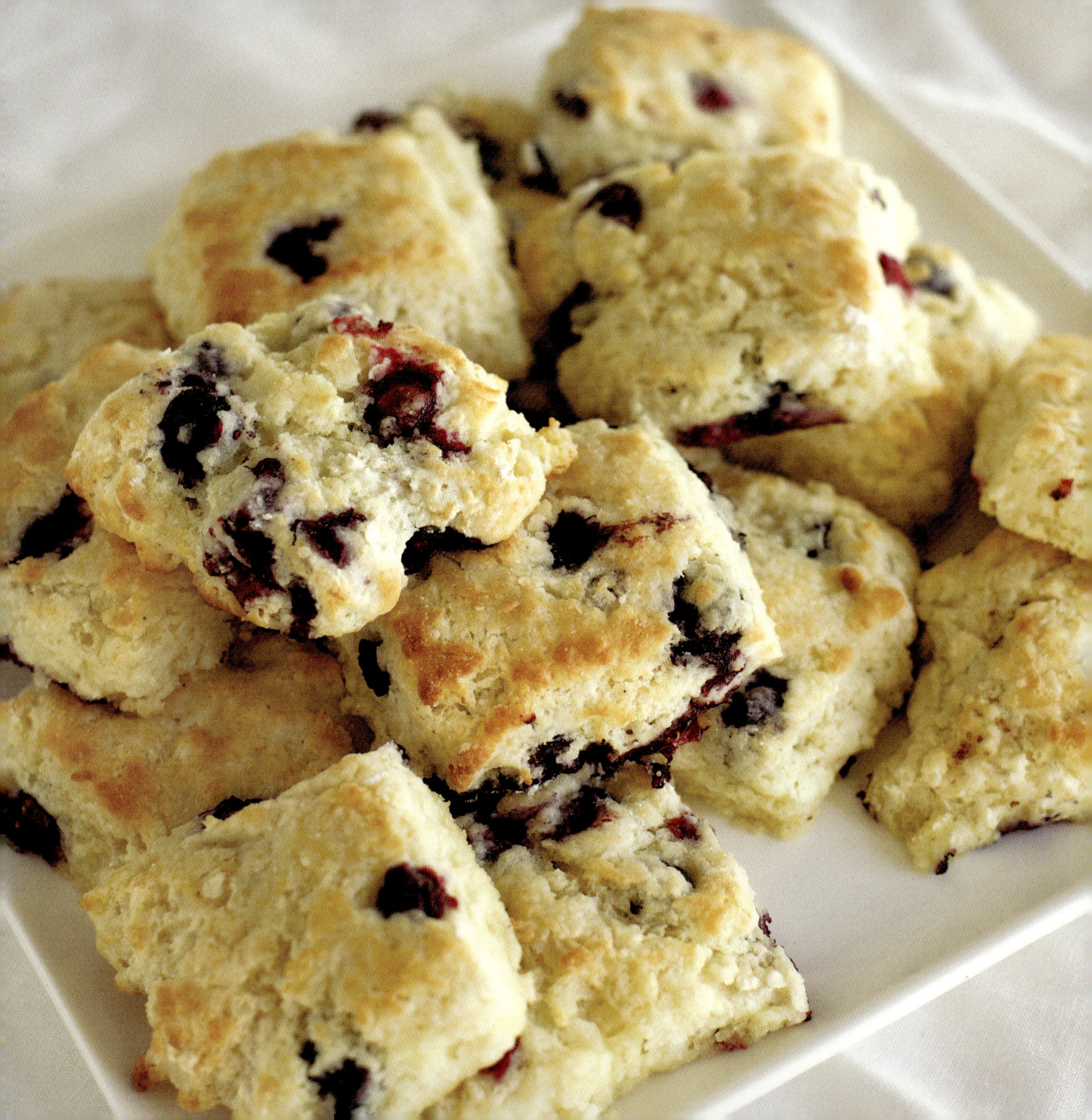

# Blueberry Sweet Biscuits

A few of you will argue that these are not really biscuits because there is added sugar, so omit the sugar, if you prefer. However, the sweeter version makes a nice base for a shortcake with fresh berries and whipped cream.

makes 16 servings

## INGREDIENTS

2¼ cups all-purpose flour
¼ cup granulated sugar
4 teaspoons baking powder
1 teaspoon salt
½ cup butter, frozen
1 cup buttermilk or whole milk
⅔ cup fresh blueberries
2 tablespoons melted butter

Preheat oven to 450°. Line a baking sheet with parchment paper, nonstick aluminum foil, or a silicone baking mat.

Combine flour, sugar, baking powder, and salt in a medium bowl. Grate butter over the large holes of a grater into flour mixture. Toss lightly until combined. Add buttermilk to flour mixture, stirring gently. Fold in blueberries.

Transfer dough to a floured surface and pat into an 8 x 8-inch square (about ½- to ¾-inch thick). Cut into 16 squares. Place biscuits on prepared pan.

Bake 8 minutes or until golden brown. Remove from oven and brush tops with melted butter.

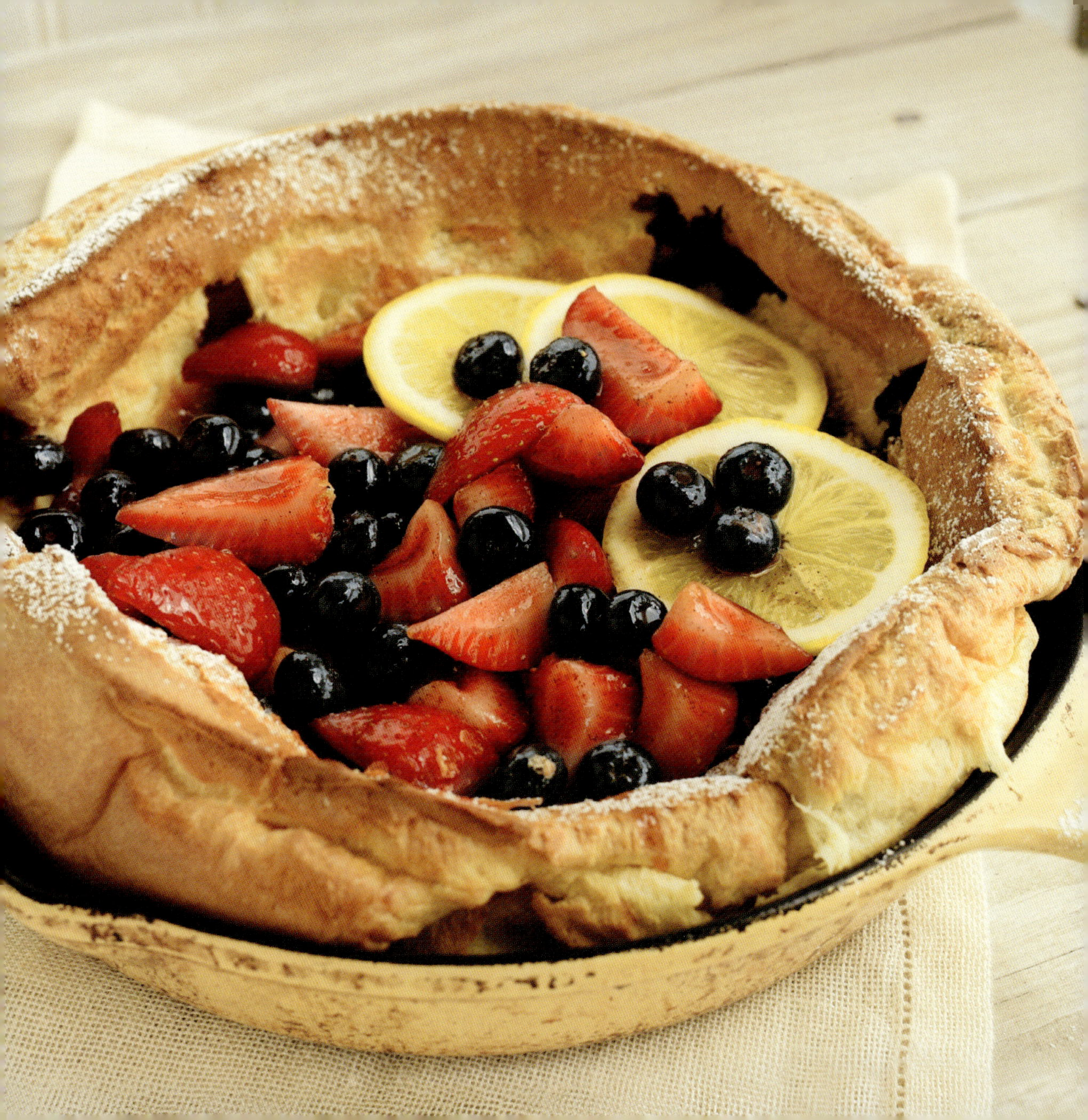

# Blueberry Dutch Baby Pancake

A Dutch Baby is basically a large, fluffy pancake that rises during baking but falls in the center when it's removed from the oven—creating the perfect spot for a luscious pile of sweetened fruit. The batter needs to be very smooth. A blender is recommended to get the correct consistency. If whisking,be sure all lumps of flour are dissolved. Frozen fruit can be used, but it needs to be thawed, and possibly drained, before spooning into the center of the hot pancake.

makes 4–6 servings

### INGREDIENTS

- 2 cups fresh or frozen and thawed blueberries, divided
- 1 cup fresh or frozen and thawed strawberries, raspberries, peaches, mango, or other fruit
- ¼ cup firmly packed light brown sugar
- ½ teaspoon ground cinnamon
- 5 large eggs
- 1¼ cups half-and-half or milk
- ½ cup melted butter
- 1 teaspoon vanilla extract
- 1¼ cups all-purpose flour
- ¼ teaspoon salt

### GARNISHES

lemon slices, powdered sugar

Preheat oven to 400°.

Combine 1 cup blueberries, strawberries, or other fruit; brown sugar; and cinnamon in a large bowl. Set aside.

Crack eggs into a blender. Add half-and-half, butter, vanilla, flour, and salt. Blend until smooth. Let stand 5 minutes.

Place a well-seasoned 12-inch cast iron skillet in oven for 5 minutes or until hot. Pour batter into hot skillet. Sprinkle remaining 1 cup blueberries on top.

Bake 20 minutes. The pancake will rise and then fall after it's removed from the oven. If desired, remove from pan and place on a serving platter. Spoon reserved sweetened fruit mixture into center and garnish, if desired.

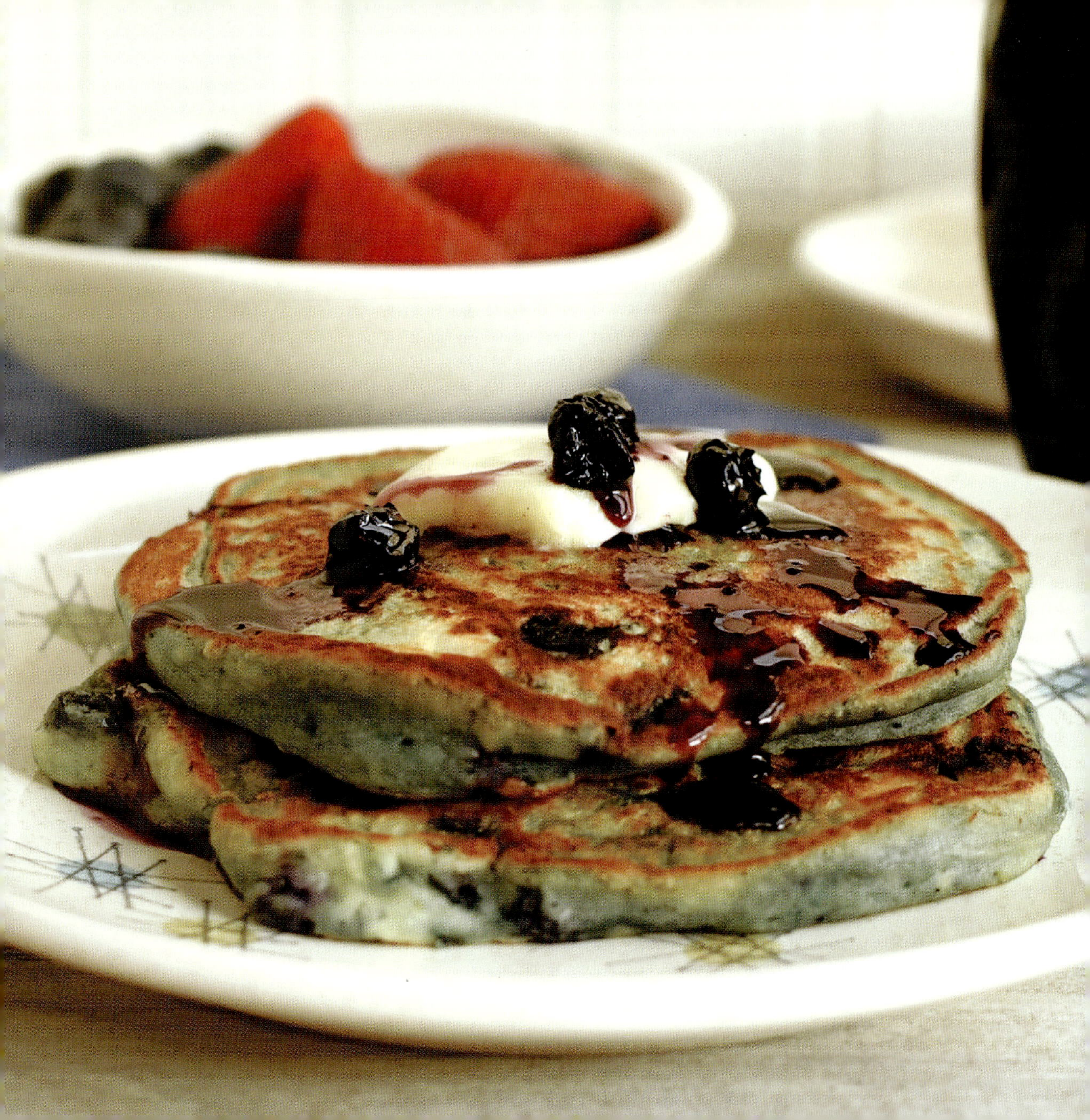

# Lemon-Blueberry Ricotta Pancakes

Using ricotta cheese keeps these pancakes moist and adds protein. You may get more than 10 pancakes out of the batter, but I usually don't count the first one or two because they are for tasting and practicing a skillet flip!

makes 10 pancakes

INGREDIENTS

1½ cups all-purpose flour
2 tablespoons granulated sugar
1 tablespoon lemon zest
2 teaspoons baking powder
½ teaspoon baking soda
½ teaspoon salt
2 large eggs
¾ cup milk
1 cup ricotta cheese
1½ tablespoons fresh lemon juice
1 cup fresh or frozen blueberries
Melted butter
Maple or Simple Blueberry Syrup (page 21)

Combine flour, sugar, lemon zest, baking powder, baking soda, and salt in a large bowl.

Whisk together eggs, milk, ricotta cheese, and lemon juice in a separate bowl. Add egg mixture to flour mixture, stirring just until combined. For tender pancakes, do not overmix. Gently fold in blueberries.

Heat a large nonstick skillet over medium heat. Brush with melted butter. Pour about ⅓ cup batter into pan for each pancake. Cook 2 minutes or until each pancake is browned on the bottom and edges appear dry. Turn and cook 1 to 2 minutes or until done in center. Repeat with remaining butter and batter. Serve with additional melted butter and syrup.

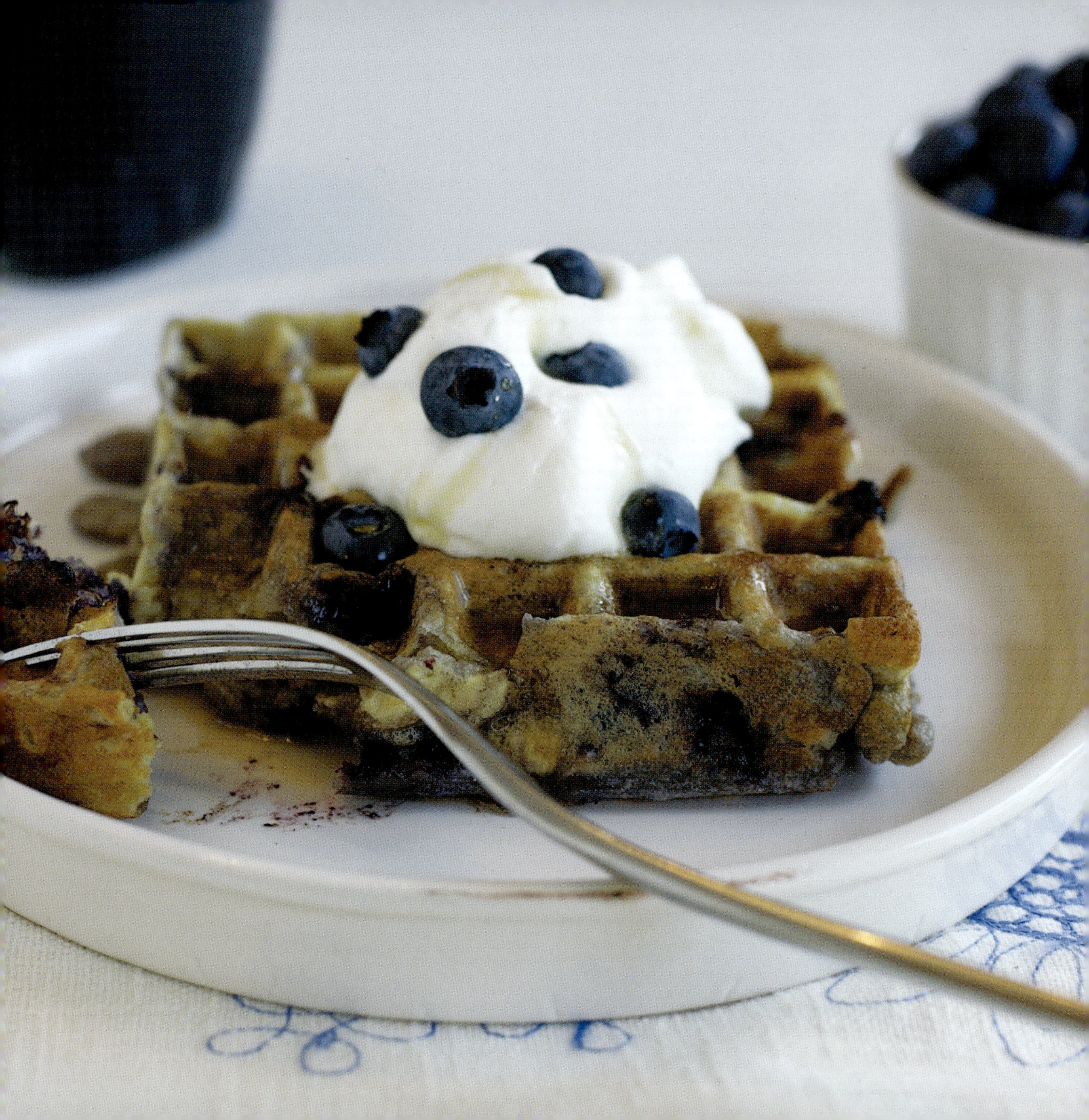

# Blueberry-Oatmeal Waffles

Oats add a nutty taste. For ease, I use thin instant oats, as old-fashioned oats can taste a bit raw. If you prefer old-fashioned oats, place them in a skillet over medium heat until toasted and a very light golden brown. Grind oats in a food processor for a smoother texture.

makes 6 servings

### INGREDIENTS

1½ cups all-purpose flour
1 cup uncooked quick or instant oats
1 tablespoon baking powder
½ teaspoon salt
2 large eggs
1½ cups buttermilk or milk
5 tablespoons butter, melted
2 tablespoons granulated sugar
1 cup fresh or frozen blueberries

### TOPPINGS

maple syrup, whipped cream, additional fresh blueberries

Combine flour, oats, baking powder, and salt in a large bowl.

Whisk eggs, buttermilk, butter, and sugar in a separate bowl. Add egg mixture to flour mixture, stirring gently until blended. Stir in blueberries. Let batter rest 5 to 10 minutes. Meanwhile, heat waffle iron according to manufacturer's directions.

Ladle recommended amount (about ½ cup for thick Belgian waffles) for each waffle into waffle iron. Cook 5 minutes or until golden brown and crisp. Repeat with remaining batter. Serve with desired toppings.

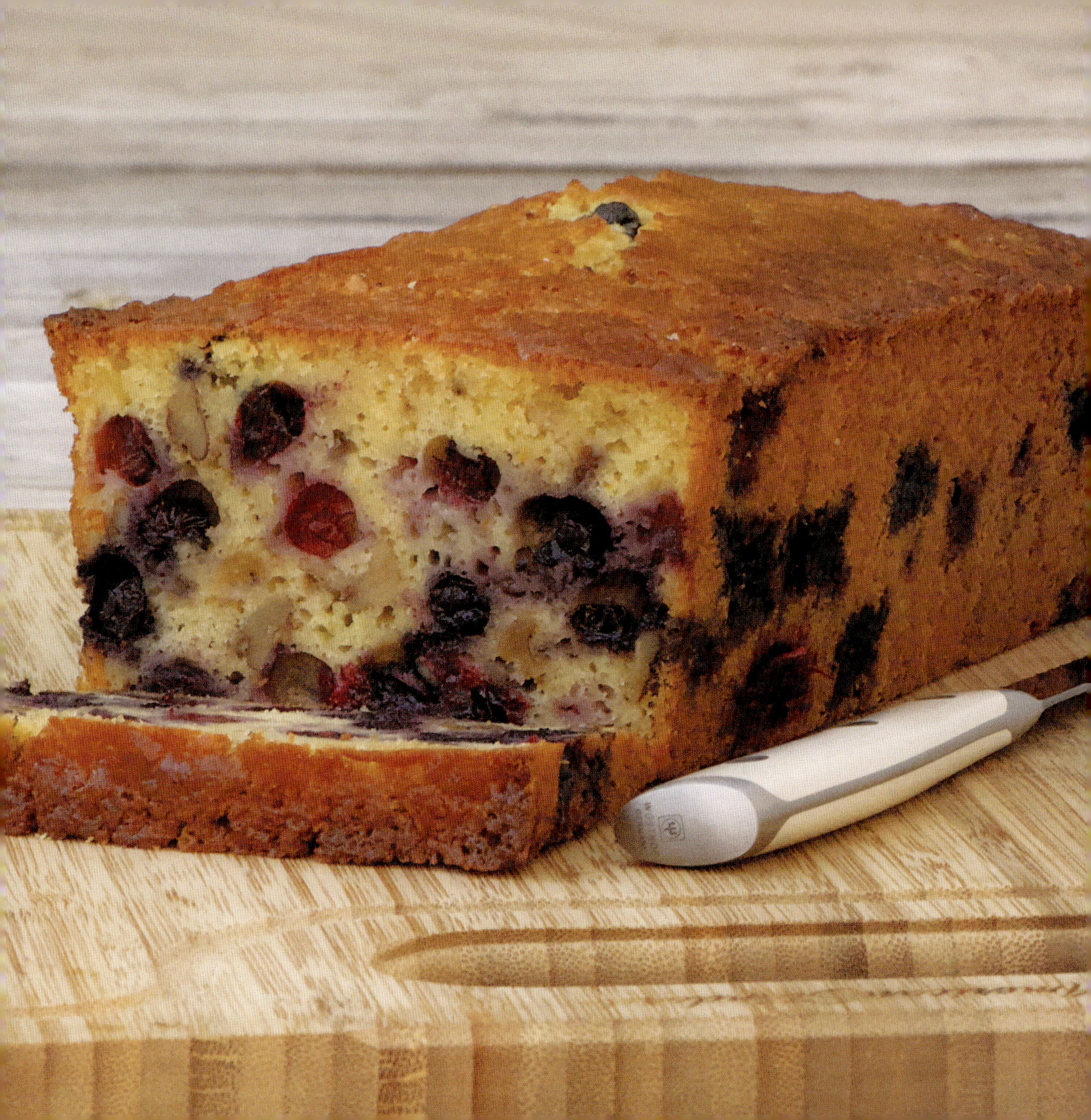

# Blueberry-Orange Bread

Blueberries pair very well with citrus. I added orange zest to this recipe; otherwise, the orange flavor might be too subtle. Try this bread toasted and spread with softened cream cheese.

makes 1 (9-inch) loaf

### INGREDIENTS

2 cups all-purpose flour
1 teaspoon baking powder
½ teaspoon salt
¾ cup butter, softened
¾ cup granulated sugar
2 large eggs
1 tablespoon orange zest
½ cup fresh orange juice
½ cup milk
1 teaspoon vanilla extract
1 cup fresh, frozen, or dried blueberries
½ cup chopped walnuts

Preheat oven to 350°. Butter a 9x5-inch loaf pan.

Combine flour, baking powder, and salt in a medium-size bowl. Set aside.

Beat butter and sugar in a large bowl with an electric mixer on medium speed until fluffy. Add eggs, one at a time, mixing well. Stir in orange zest, orange juice, milk, and vanilla.

Stir flour mixture into butter mixture, just until dry ingredients are moistened. (Do not overmix). Fold in berries and walnuts. Spoon batter into prepared pan.

Bake for 55 to 60 minutes or until center is dry when tested with a wooden pick (add 5 to 10 minutes if using frozen berries). Cool in pan 10 minutes; remove loaf and transfer to a wire rack to cool.

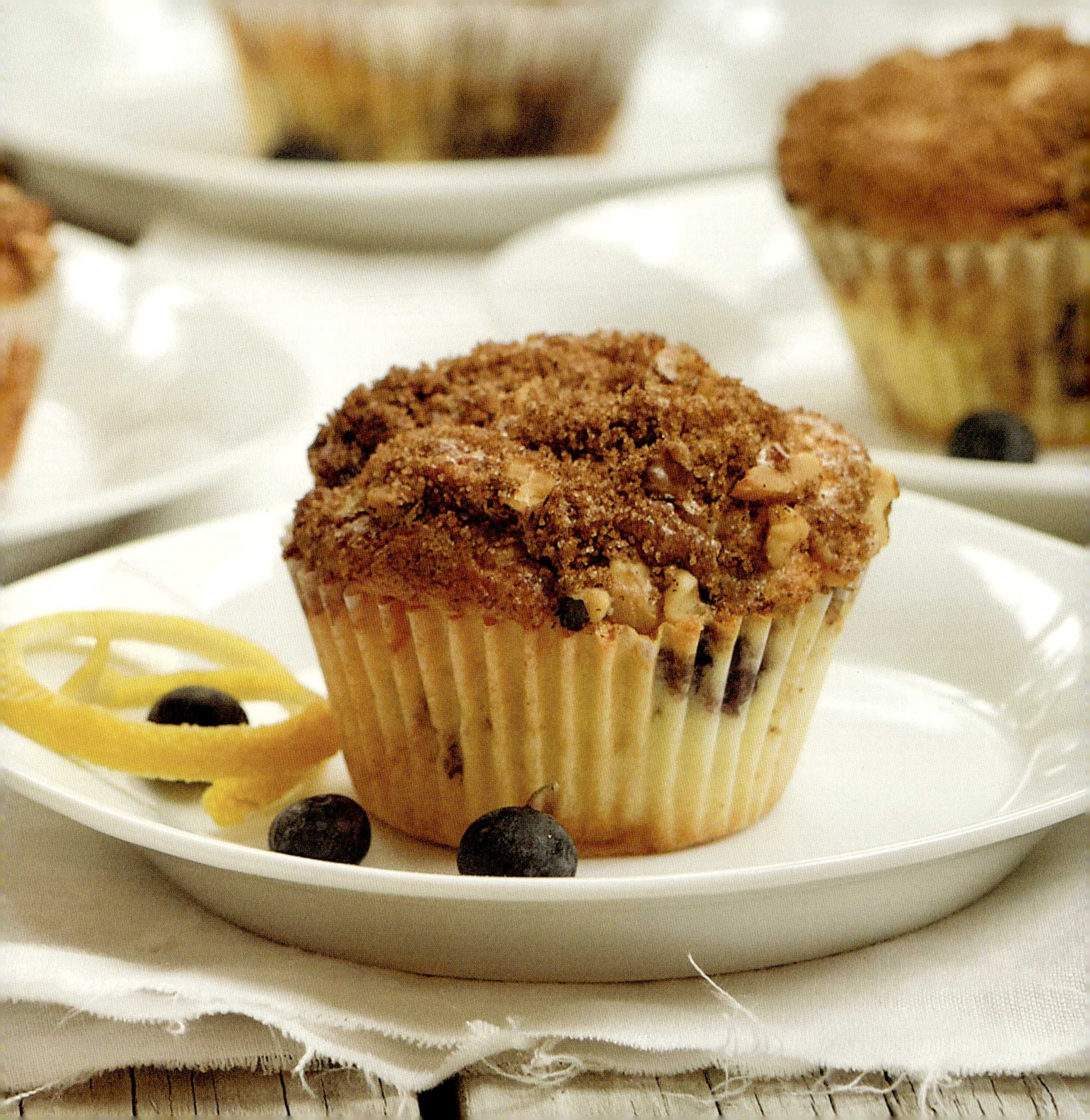

# Blueberry-Lemon Crunch Muffins

Cecilia Glembocki from the Virginia Egg Council gave me the recipe for these delicious muffins—some of the best you'll ever try. The crunchy nut topping takes them from simple to simply sublime. These freeze very well, so consider doubling the recipe to have a quick snack or breakfast muffin anytime.

makes 1 dozen

INGREDIENTS

Crunch Topping (recipe at right)
½ cup butter, softened
½ cup granulated sugar
1 tablespoon lemon zest
3 eggs, separated
¼ cup fresh lemon juice
1 cup all-purpose flour
1 teaspoon baking powder
¼ teaspoon salt
1 cup fresh or frozen blueberries

Preheat oven to 375°. Place 12 muffin cup liners in a muffin pan, or grease muffin pan with butter; set aside.

Prepare Crunch Topping; set aside.

Beat butter, sugar, and lemon zest in a large bowl with an electric mixer until creamy. Beat in egg yolks and lemon juice.

Stir together flour, baking powder, and salt in a separate bowl. Add flour mixture to butter mixture, stirring just until moistened.

Beat egg whites until stiff peaks form. Fold egg whites and blueberries into batter. Spoon batter into prepared muffin cups, filling ¾ full.

Sprinkle each muffin cup with 1 tablespoon Crunch Topping. Bake for 20 to 25 minutes. Cool in pan on a wire rack 5 minutes; then remove to a wire rack to cool completely.

*Crunch Topping:* Combine **⅓ cup chopped walnuts, ⅓ cup firmly packed light brown sugar,** and **1½ teaspoons ground cinnamon** in a small bowl, stirring until well blended.

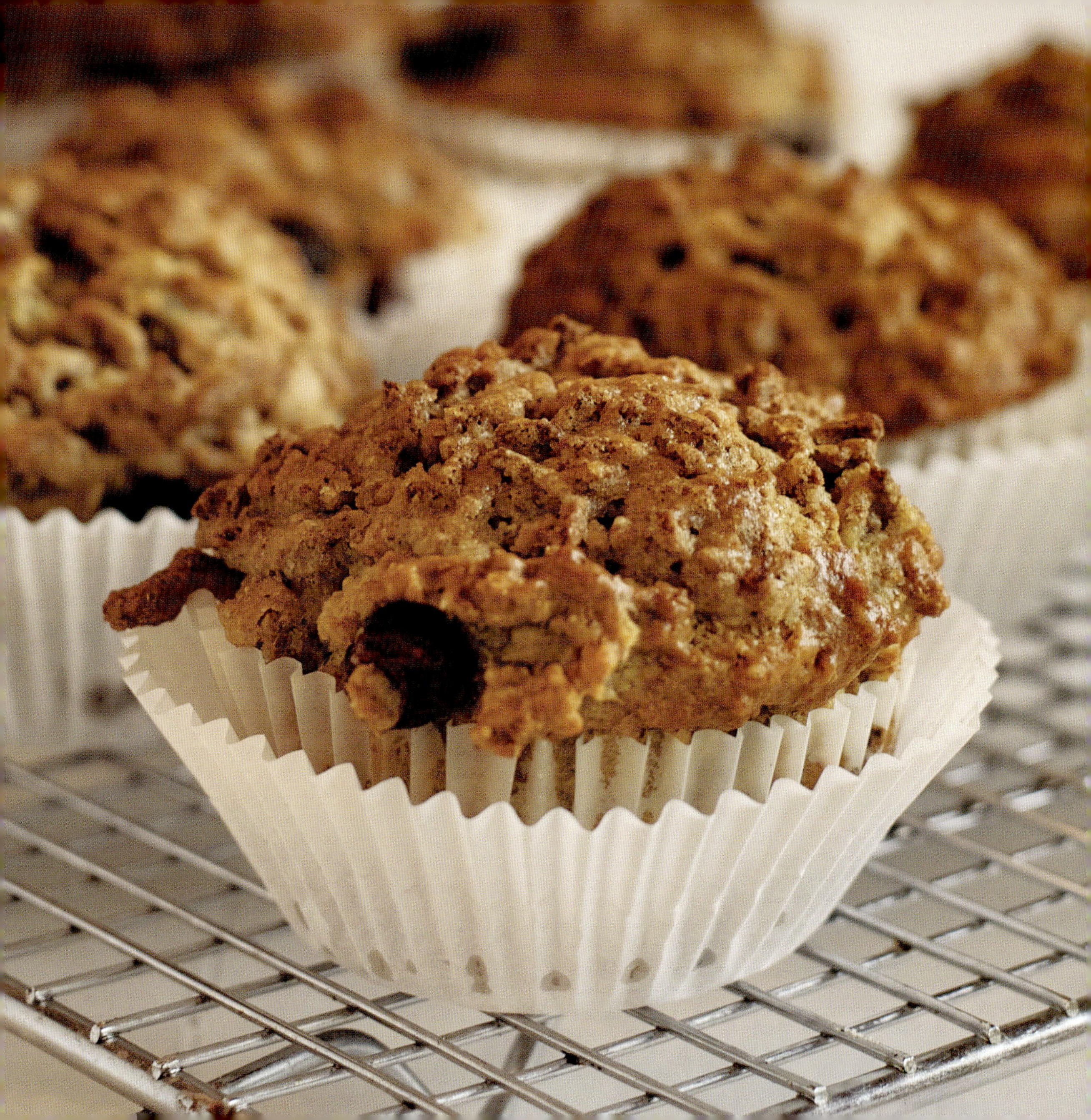

# Rise 'n' Shine Blueberry Bran Muffins

My dear friend Susan Dosier and I cowrote a book called *Discover Dinnertime: Your Guide to Building Family Time Around the Table* years ago when our girls were quite young. Susan created this yummy, filling breakfast muffin based off of one that we had tested when we worked together at *Southern Living* magazine. The original recipe used skim milk and substituted applesauce for oil to make it low in fat. Nowadays, we're all about "good fat," and I use whole milk and organic canola oil in this version.

makes 1 dozen

INGREDIENTS

Buttery Bran-Streusel Topping (recipe at right)
1 large egg
1¼ cups whole milk
¼ cup canola oil
2 cups shredded bran cereal (such as All-Bran)
1¼ cups all-purpose flour
½ cup firmly packed light brown sugar
4 teaspoons baking powder
½ teaspoon salt
1 cup fresh or frozen blueberries
1 teaspoon vanilla extract

Preheat oven to 400°. Place 12 muffin cup liners in a muffin pan, or lightly grease muffin pan with cooking spray.

Prepare Buttery Bran-Streusel Topping, and set aside.

Whisk egg, milk, and oil in a large bowl. Stir in cereal. Let stand 10 minutes to allow cereal to soften.

Combine flour, brown sugar, baking powder, and salt in a separate bowl. Add flour mixture to cereal mixture, stirring to combine (do not overmix). Stir in blueberries and vanilla.

Spoon batter evenly into prepared muffin cups, filling ¾ full. Sprinkle evenly with topping. Bake for 20 to 25 minutes. Let stand in pans 1 minute; remove to a wire rack to cool completely.

*Buttery Bran-Streusel Topping:* Combine **2 tablespoons all-purpose flour, 3 tablespoons firmly packed light brown sugar,** and **¼ cup shredded bran cereal** in a small bowl. Cut in **½ cup butter** with a fork or pastry blender until crumbly.

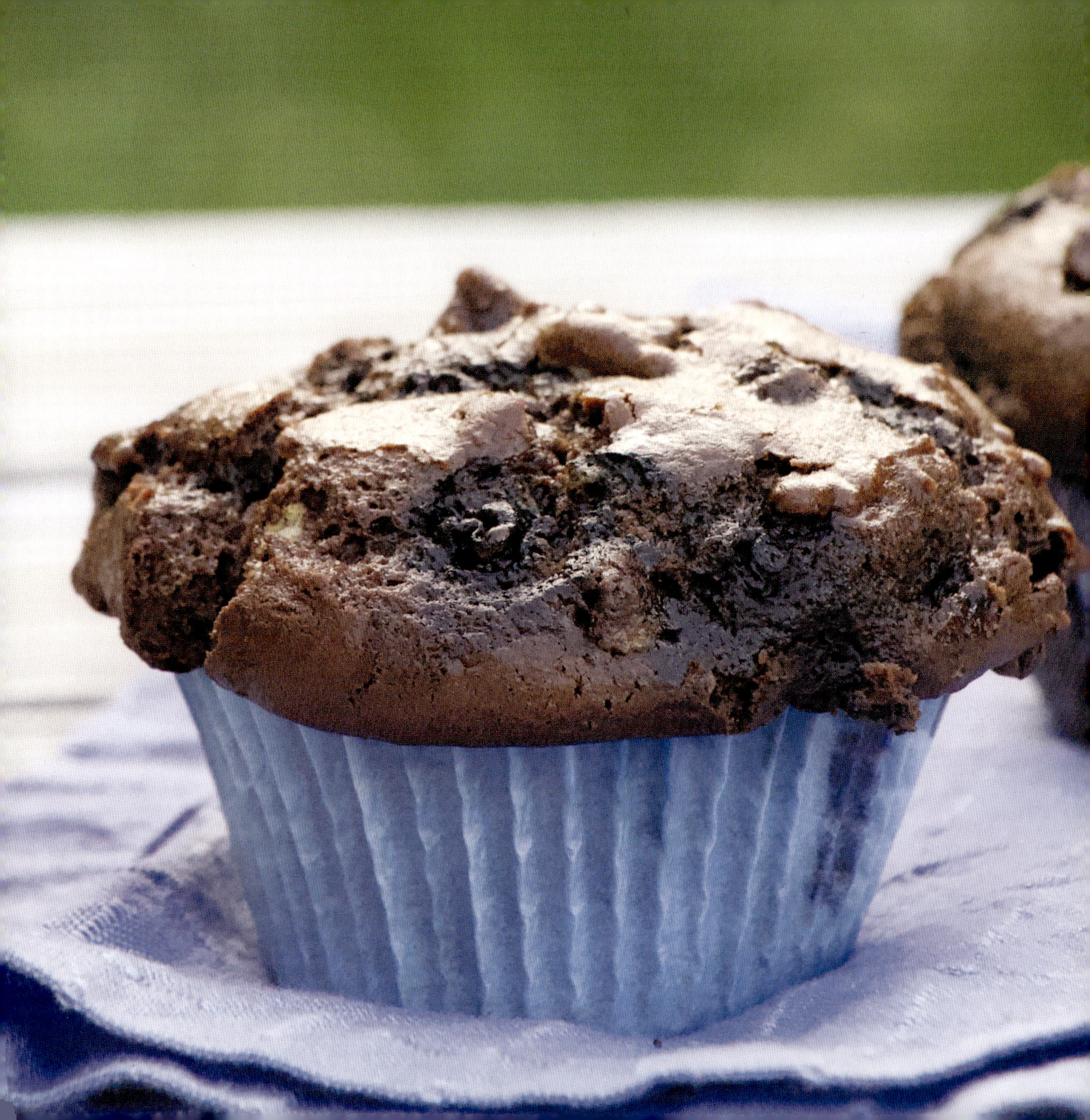

# Double Chocolate Blueberry Muffins

This is now my favorite muffin. The dense chocolate flavor is brownie-like, and the blueberries and chocolate chips ensure a super-moist texture. If you can resist eating them all at once, freeze some for future on-the-go breakfasts.

makes 1 dozen

## INGREDIENTS

- 2 cups plus 1 tablespoon all-purpose flour, divided
- 1 cup cocoa powder
- 2 teaspoons baking powder
- ¾ teaspoon baking soda
- ½ teaspoon salt
- 1½ cups fresh or frozen blueberries
- 3 large eggs
- ¾ cup granulated sugar
- ½ cup firmly packed light brown sugar
- ½ cup vegetable oil
- 1 cup sour cream
- 2 teaspoons vanilla extract
- 1 cup semisweet or dark chocolate chips

Preheat oven to 350°. Place 12 muffin cup liners in a muffin pan.

Combine 2 cups flour, cocoa powder, baking powder, baking soda, and salt in a large bowl. Combine blueberries and remaining 1 tablespoon flour in a small bowl.

Whisk eggs and sugars together in a separate large bowl until sugars almost dissolve. Whisk in oil, sour cream, and vanilla. Stir egg mixture into flour mixture, stirring just until dry ingredients are moistened. Fold in chocolate chips and blueberries. Spoon batter into prepared muffin cups, filling ¾ full.

Bake 25 minutes or until center is dry when tested with a wooden pick (add 5 to 10 minutes if using frozen berries; chocolate chips and blueberries will remain wet.) Cool in pan 15 minutes; then remove to a wire rack to cool completely.

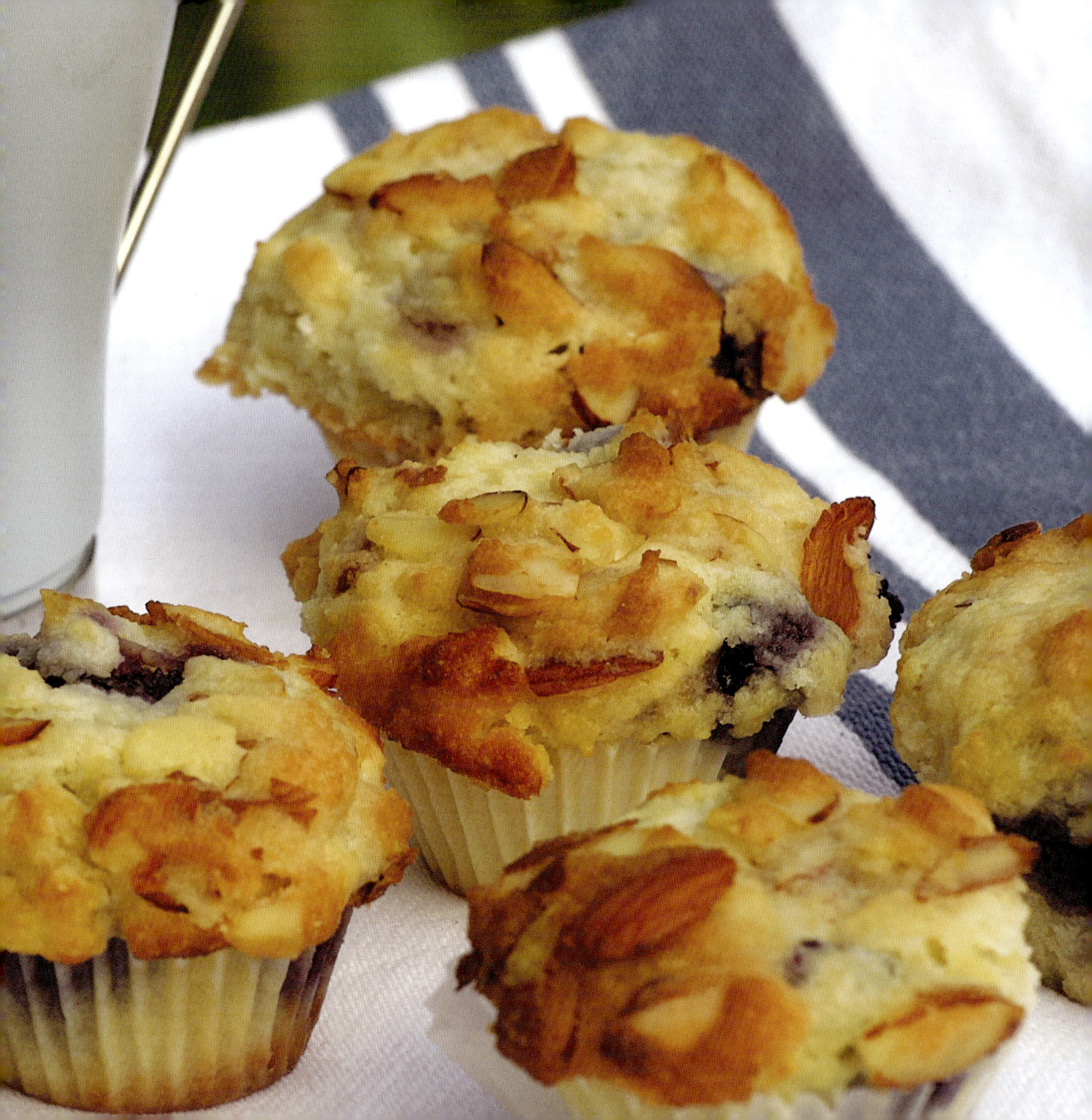

# Keto-Friendly Almond Streusel Blueberry Mini Muffins

Almond and coconut flours are gluten-free substitutes for regular wheat flour, and this recipe does a good job of mimicking the texture of traditional muffins. If you are not a fan of almond extract, vanilla extract may be substituted.

makes 2 dozen

INGREDIENTS

Almond Streusel Topping (recipe at right)
3 large eggs
⅓ cup coconut oil or melted butter
⅓ cup unsweetened almond milk or whole milk
1 teaspoon almond extract
2½ cups almond flour or coconut flour
½ cup Erythritol sugar substitute or granulated sugar
1½ teaspoons baking powder
¼ teaspoon salt
1 cup fresh or frozen blueberries

Preheat oven to 350°. Lightly grease 2 (12-cup) mini-muffin pans with cooking spray.

Prepare Almond Streusel Topping; set aside.

Whisk together eggs in a large bowl. Stir in oil, milk, and almond extract. Combine flour, sugar, baking powder, and salt in a separate bowl. Stir flour mixture into egg mixture, just until moistened. Fold in blueberries.

Spoon batter evenly into prepared muffin cups, filling ¾ full. Sprinkle evenly with Almond Streusel Topping. Bake 20 to 25 minutes.

*Almond Streusel Topping:* Combine **⅓ cup almond or coconut flour** and **⅓ cup Erythritol sugar substitute or granulated sugar** in a bowl. Add **4 tablespoons softened butter,** stirring until combined. Stir in **⅓ cup sliced almonds.**

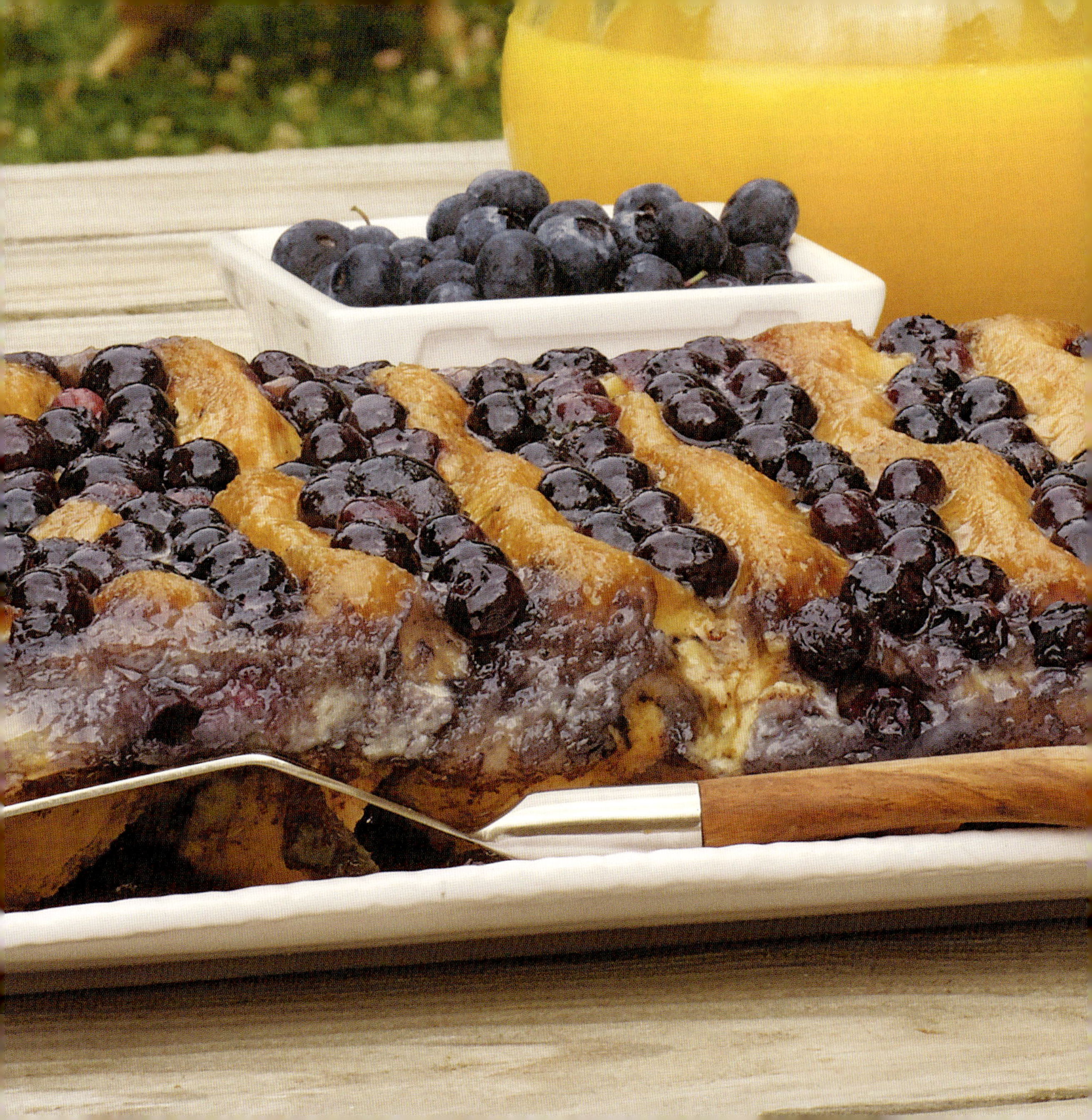

# Upside-down Blueberry French Toast

This is definitely a breakfast fit for company! Conveniently, it's best when made the night before, so the bread can soak up the liquid. You might be tempted to lift away the individual bread slices when serving, but it's easier to cut it into squares. Make sure your platter has sides so the syrupy liquid stays contained.

makes 8–12 servings

INGREDIENTS

¾ cup firmly packed light brown sugar
½ cup melted butter
¼ cup maple syrup
2 cups fresh or frozen blueberries
1 (16-ounce) loaf French or Italian bread
8 large eggs
2½ cups half-and-half or milk
½ cup granulated sugar
1 teaspoon ground cinnamon
1 teaspoon ground nutmeg
2 teaspoons vanilla extract
⅓ cup chopped pecans, walnuts, or almonds (optional)

Line a 13x9-inch baking dish with nonstick aluminum foil or with foil lightly greased with butter.

Combine brown sugar, melted butter, and syrup in a small bowl, stirring until well blended. Pour brown sugar mixture into bottom of prepared baking dish. Sprinkle with blueberries. Slice bread into 16 (1-inch) slices; arrange over brown sugar mixture, overlapping to fit.

Whisk eggs in a bowl. Stir in half-and-half, granulated sugar, cinnamon, nutmeg, and vanilla. Pour half-and-half mixture over bread, pressing down until bread is soaked. Cover with plastic wrap and refrigerate 6 to 8 hours or overnight.

Preheat oven to 350°. Uncover and sprinkle with nuts, if desired. Cover with foil and bake 25 minutes. Uncover and cook 10 more minutes or until golden brown. Let stand 10 minutes.

Place a heatproof serving tray over pan; invert dish and tray. Remove dish and foil. Or cut into individual pieces and turn upside-down onto plates, scooping sauce and berries with a spoon and drizzling over top of each serving.

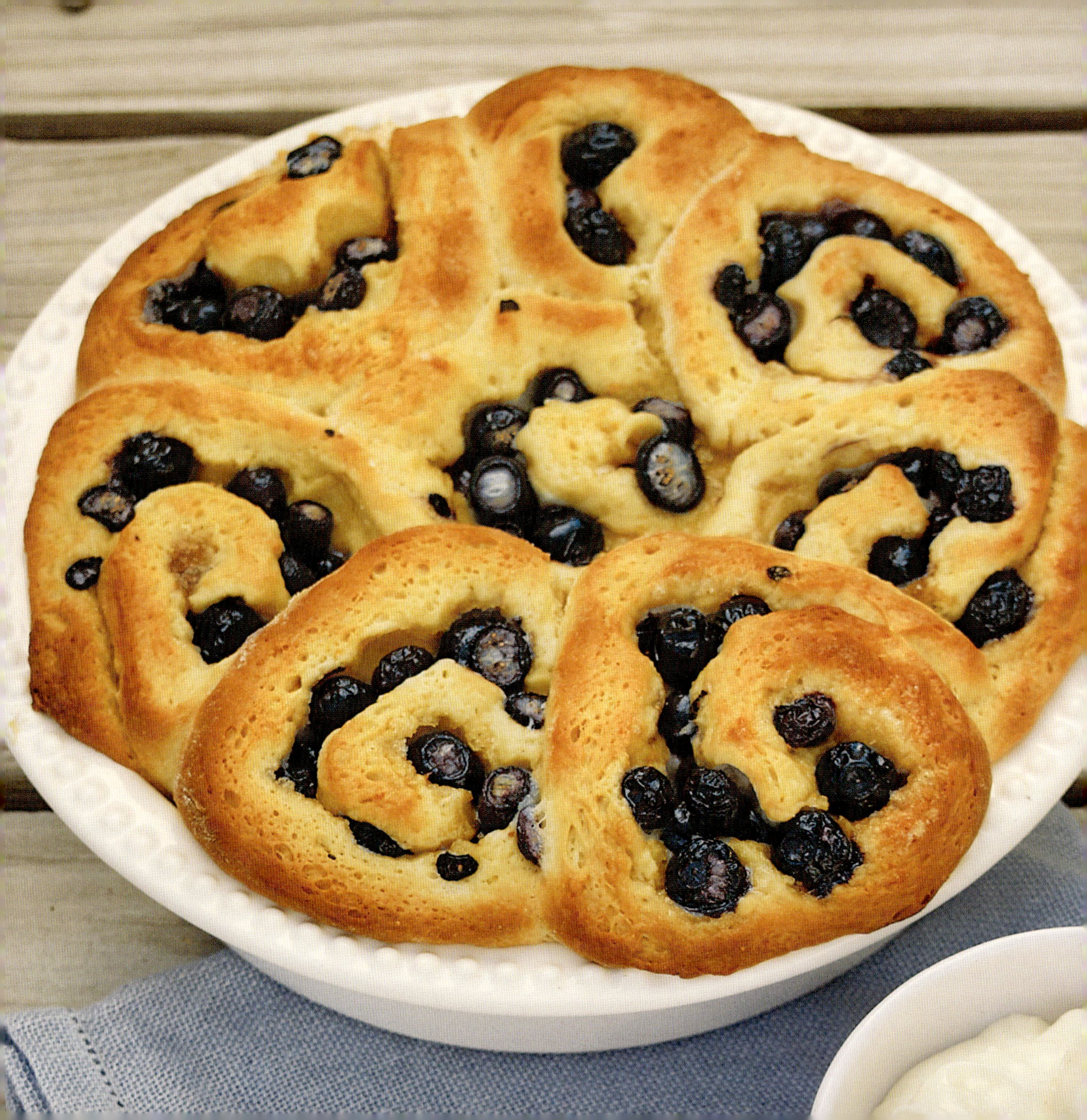

# Blueberry-Orange Sweet Rolls

Yum, yum, yum. Enough said.

makes 1 dozen

INGREDIENTS

1 (¼-ounce) envelope active dry yeast (2¼ teaspoons)
½ cup warm orange juice or water (100° to 110°)
⅓ cup melted butter
⅓ cup granulated sugar
2 large eggs
1 teaspoon orange zest
3 cups bread flour or all-purpose flour, divided
1 teaspoon salt
3 tablespoons butter, softened
⅓ cup firmly packed light brown sugar
2 cups fresh or frozen and thawed blueberries
Orange Icing (recipe at right)

Combine yeast in ½ cup warm juice or water in a mixing bowl; let stand 5 minutes. Add melted butter, granulated sugar, eggs, and zest; beat with a mixer at medium speed until blended. Add 2 cups flour and salt; beat until smooth. Gradually add remaining 1 cup flour, stirring to a soft dough.

Place dough in a large, lightly greased bowl, turning to coat. Cover and let rise in a warm place (85°), free from drafts, 1 hour or until doubled in size.

Punch dough down; turn out onto a lightly floured surface. Roll dough into a 9x18-inch rectangle. Spread softened butter over dough, leaving a ½-inch border. Sprinkle evenly with brown sugar and blueberries. Gently roll up dough, starting at long side. Cut into 12 slices, about 1½ inches thick. Place slices in a lightly greased 9x9-inch or 2-quart baking dish. Cover and let rise 20 to 30 minutes or until doubled in size.

Preheat oven to 350°. Uncover dough; bake 25 minutes or until golden. Drizzle icing over warm rolls. Makes 1 dozen.

*Orange Icing:* Beat **2 tablespoons butter** and **4 ounces softened cream cheese** in a bowl on medium speed. Gradually add **2 cups sifted powdered sugar**, mixing on low speed until smooth. Beat in **2 teaspoons fresh orange juice** and **1 teaspoon orange zest.** Increase speed and beat until smooth.

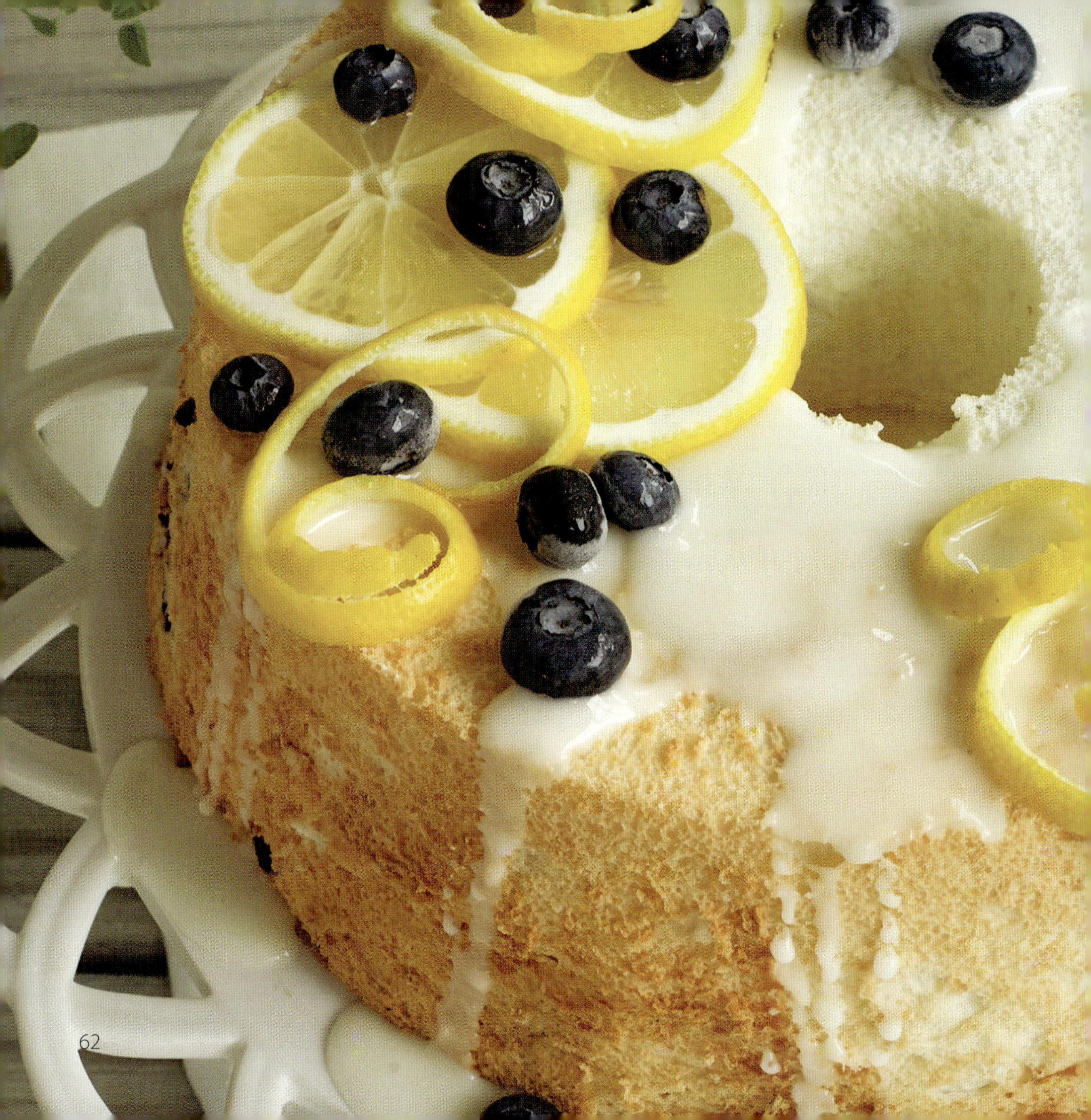

# cookies and cakes

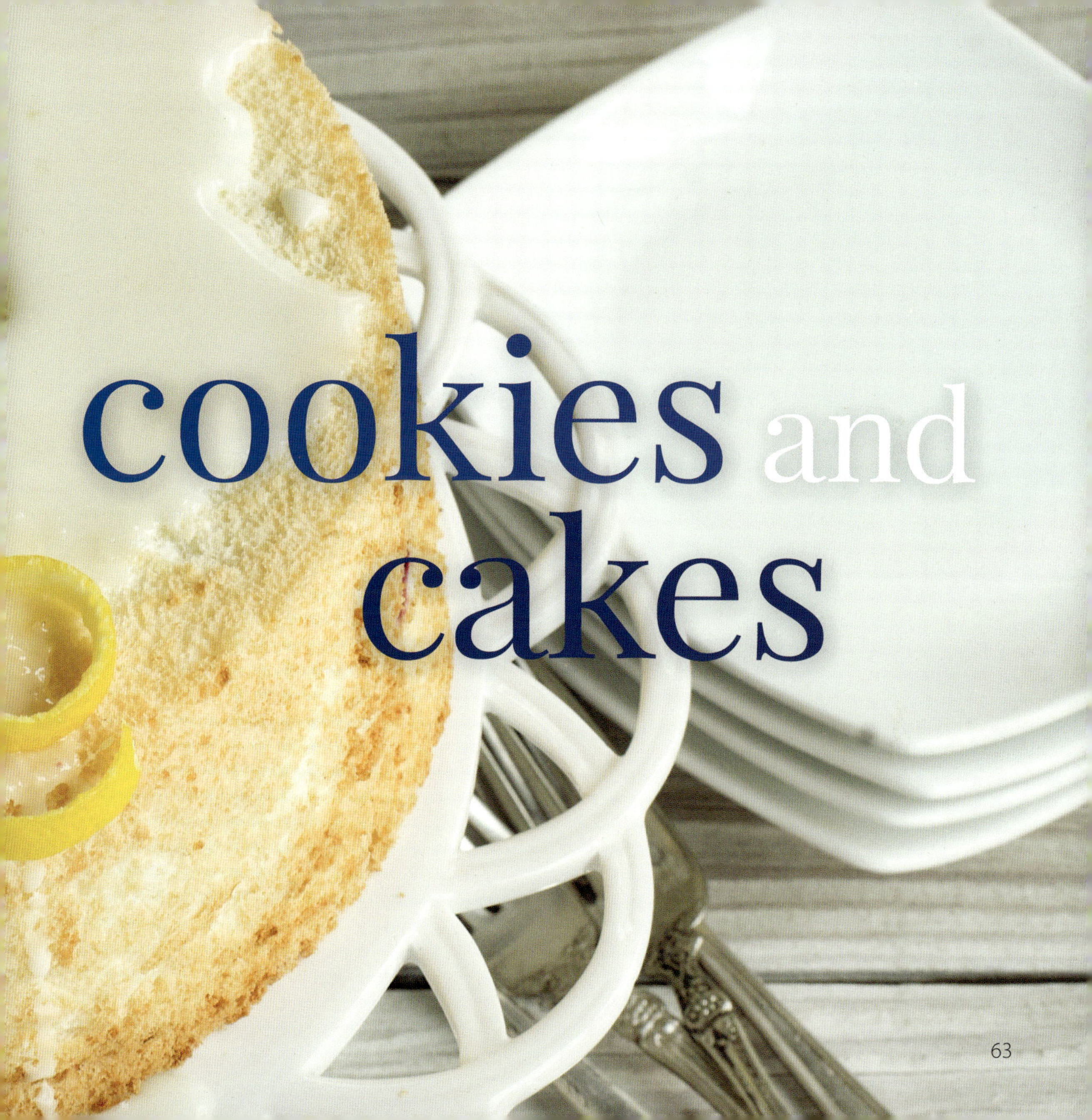

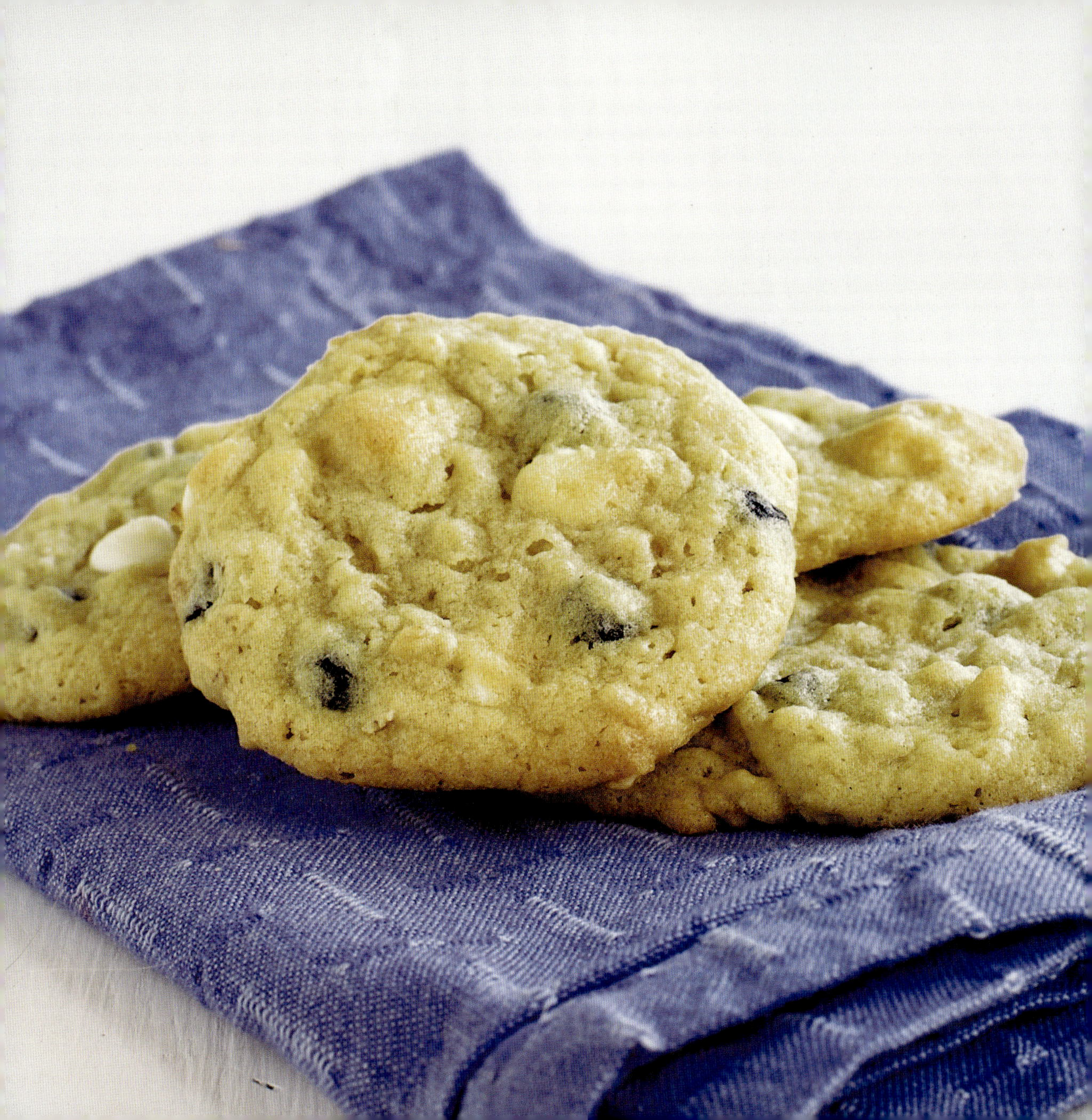

# Blueberry-White Chocolate Chip Cookies

I like using dried blueberries in this simple cookie when I need a shelf-stable treat that will keep for a few days in an airtight container. Fresh berries certainly work and provide an interesting, lumpy texture, but cookies made with fresh berries should be refrigerated. For a toothier cookie, stir in ½ cup quick-cooking oats.

makes 4 dozen

INGREDIENTS

1 cup butter, softened
1 cup firmly packed light brown sugar
½ cup granulated sugar
2 large eggs
2 teaspoons vanilla extract
2½ cups all-purpose flour
1 teaspoon baking soda
1 teaspoon baking powder
½ teaspoon salt
1 (12-ounce) package white chocolate chips
1 cup dried blueberries or 1½ cups fresh blueberries

Preheat oven to 350°. Butter a baking sheet or line with parchment paper, nonstick aluminum foil, or a silicone baking mat.

Beat butter and sugars in a large bowl at medium speed with an electric mixer until creamy. Add eggs, one at a time, beating until blended after each addition. Beat in vanilla.

Combine flour, baking soda, baking powder, and salt in a separate bowl. Slowly add flour mixture to butter mixture, beating until blended. Stir in chocolate chips and blueberries. Drop dough by heaping teaspoonfuls onto baking sheets.

Bake for 10 to 12 minutes. Cool on pan 5 minutes; transfer to a wire rack to cool completely.

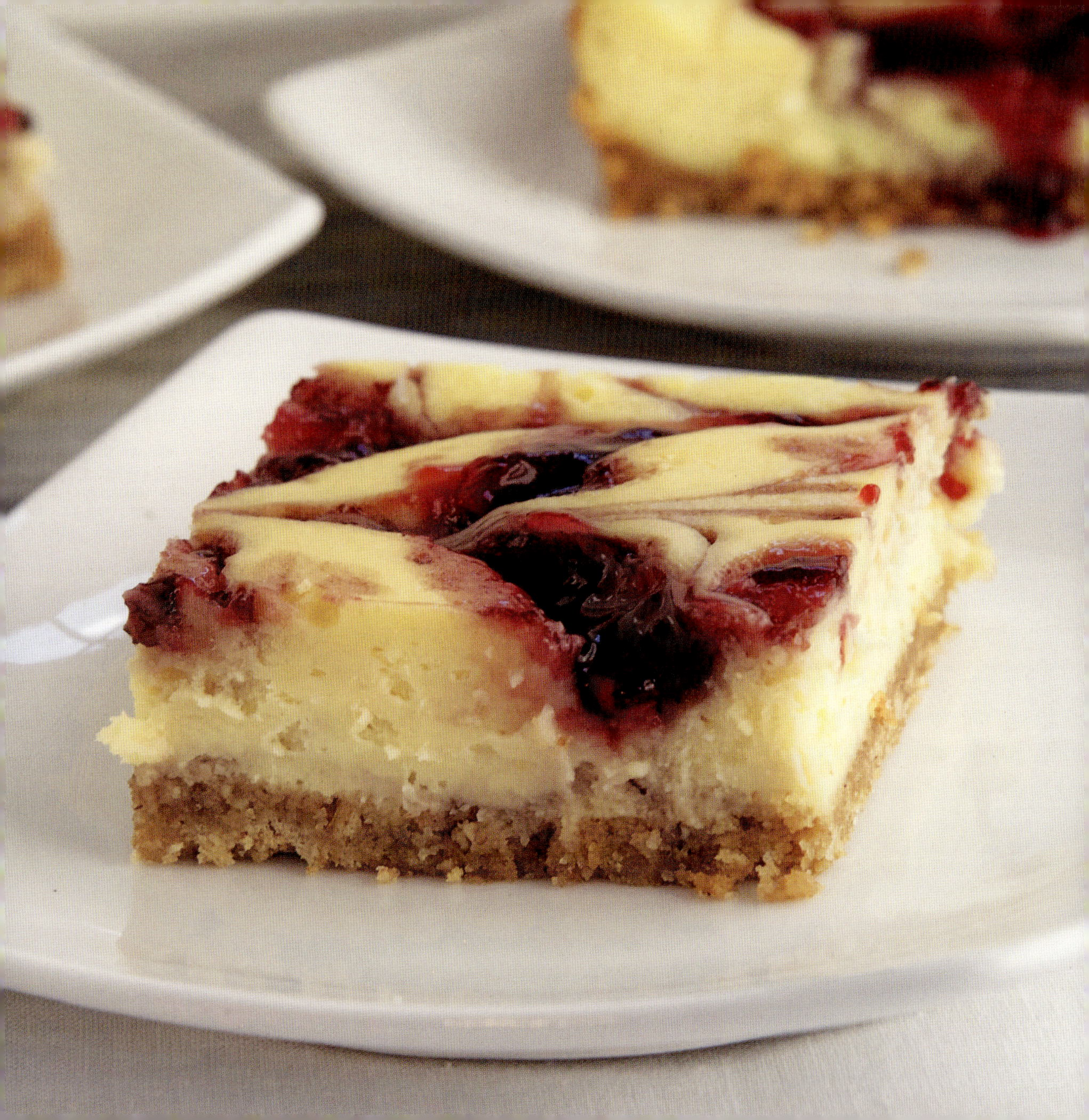

# Blueberry Cheesecake Bars

These are great make-ahead sweet treats. Cut them large for a regular dessert or into bite-size squares to serve a crowd.

makes 2 dozen

## INGREDIENTS

2 cups graham cracker crumbs
1⅓ cups granulated sugar, divided
½ cup melted butter
3 (8-ounce) packages cream cheese, softened
¼ cup sour cream
2 tablespoons all-purpose flour
3 large eggs
1 tablespoon vanilla extract
1 cup Simplest Blueberry Jam (page 23 or store-bought) or blueberry preserves

Preheat oven to 350°. Line a 13x9-inch baking dish with nonstick aluminum foil or lightly buttered foil.

Combine graham cracker crumbs, ⅓ cup sugar, and butter in a medium-size bowl, stirring until well blended. Press graham cracker mixture into prepared baking dish.

Bake 10 minutes or until light golden but not browned. Set aside on a wire rack to cool. Reduce oven temperature to 325°.

Beat cream cheese, sour cream, and remaining 1 cup sugar in a large bowl with an electric mixer until creamy. Beat in flour, eggs, and vanilla. Spread batter over graham crust.

Whisk jam until spreadable. (If preserves are too thick, whisk in 1 teaspoon water.) Drop teaspoons of preserves over top of batter, and swirl with a knife or spatula.

Bake 35 minutes or until mixture is almost set in center. Center will move slightly but won't be liquid. Cool on a wire rack. Chill until firm; cut into squares.

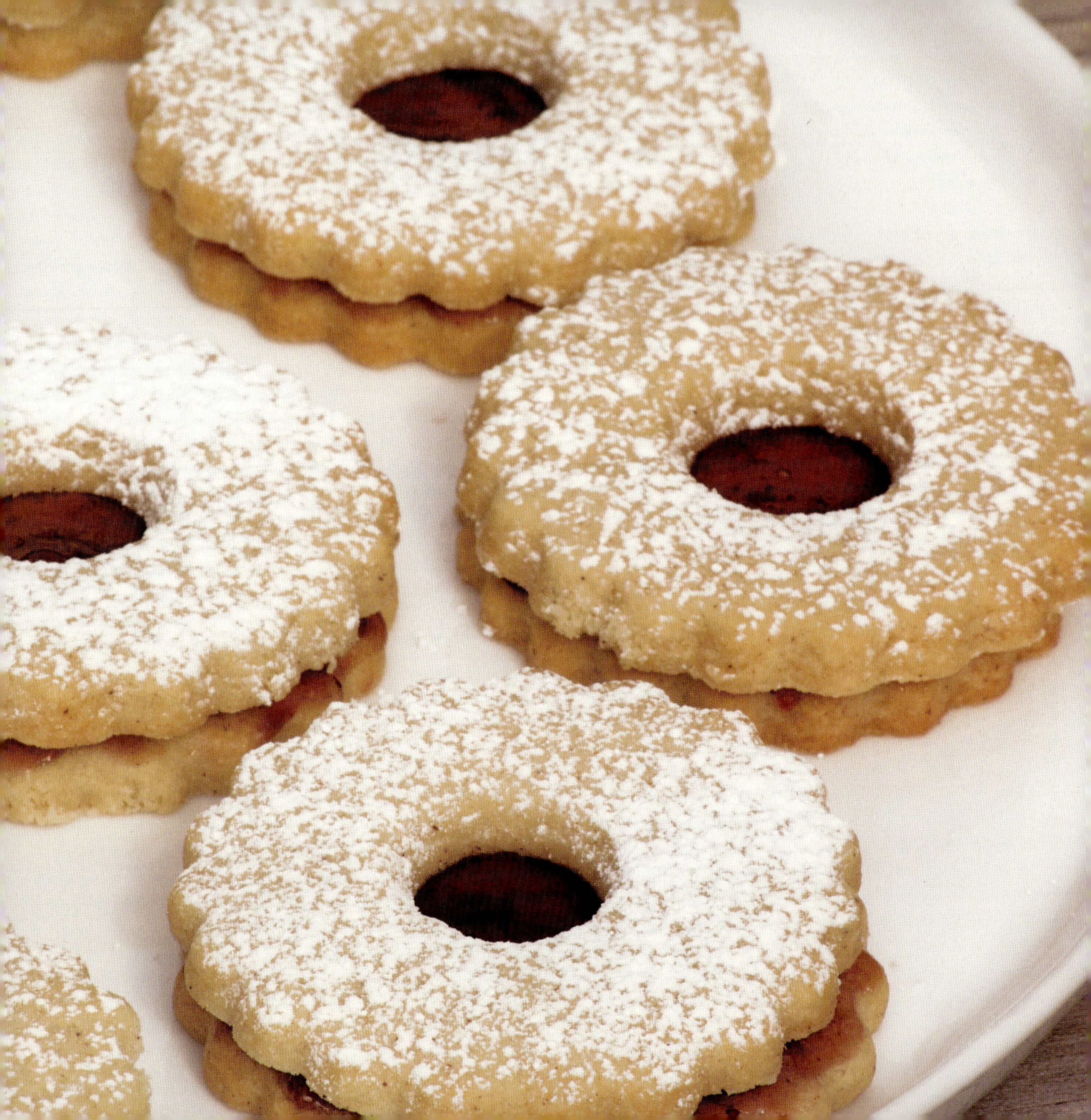

# Blueberry Linzer Cookies

Personalize these sweet bites by using your favorite shaped cookie cutter and cutting the inside with another smaller shape. Have fun with it (think dinosaur with a tiny heart center!), but remember that larger cookies will take a minute or so longer to bake—and change the yield.

makes 20

INGREDIENTS

1 cup butter, at room temperature
¾ cup granulated sugar
1 large egg
1 teaspoon vanilla extract
1¾ cups all-purpose flour
1 cup hazelnut meal or almond flour
¼ teaspoon salt
½ teaspoon ground cinnamon
Powdered sugar
½ cup Simplest Blueberry Jam (page 23 or store-bought)

Beat butter and granulated sugar in a large bowl with an electric mixer until creamy. Beat in egg and vanilla.

Stir together flour, hazelnut meal, salt, and cinnamon in a separate bowl. Add flour mixture to egg mixture, stirring just until combined.

Divide dough in half and press each half into a thin disk. Cover in plastic wrap and chill several hours or until firm.

Preheat oven to 350°. Line a baking sheet with parchment paper, nonstick aluminum foil, or a silicone baking mat.

Roll dough ¼-inch thick on a lightly floured surface. Cut into 2½-inch rounds. Cut a circle or other shape out of center of half the cookies. Place the cut-out pieces on empty spaces of baking sheet and bake with regular cookies. (Reroll dough scraps and make more cookies.)

Bake 8 to 10 minutes or until edges just start to brown. Cool on pan 5 minutes; transfer to a wire rack to cool completely.

Dust cookies with holes in center with powdered sugar. Spread 1 teaspoon jam on solid cookies. Top with sugar-dusted cookies.

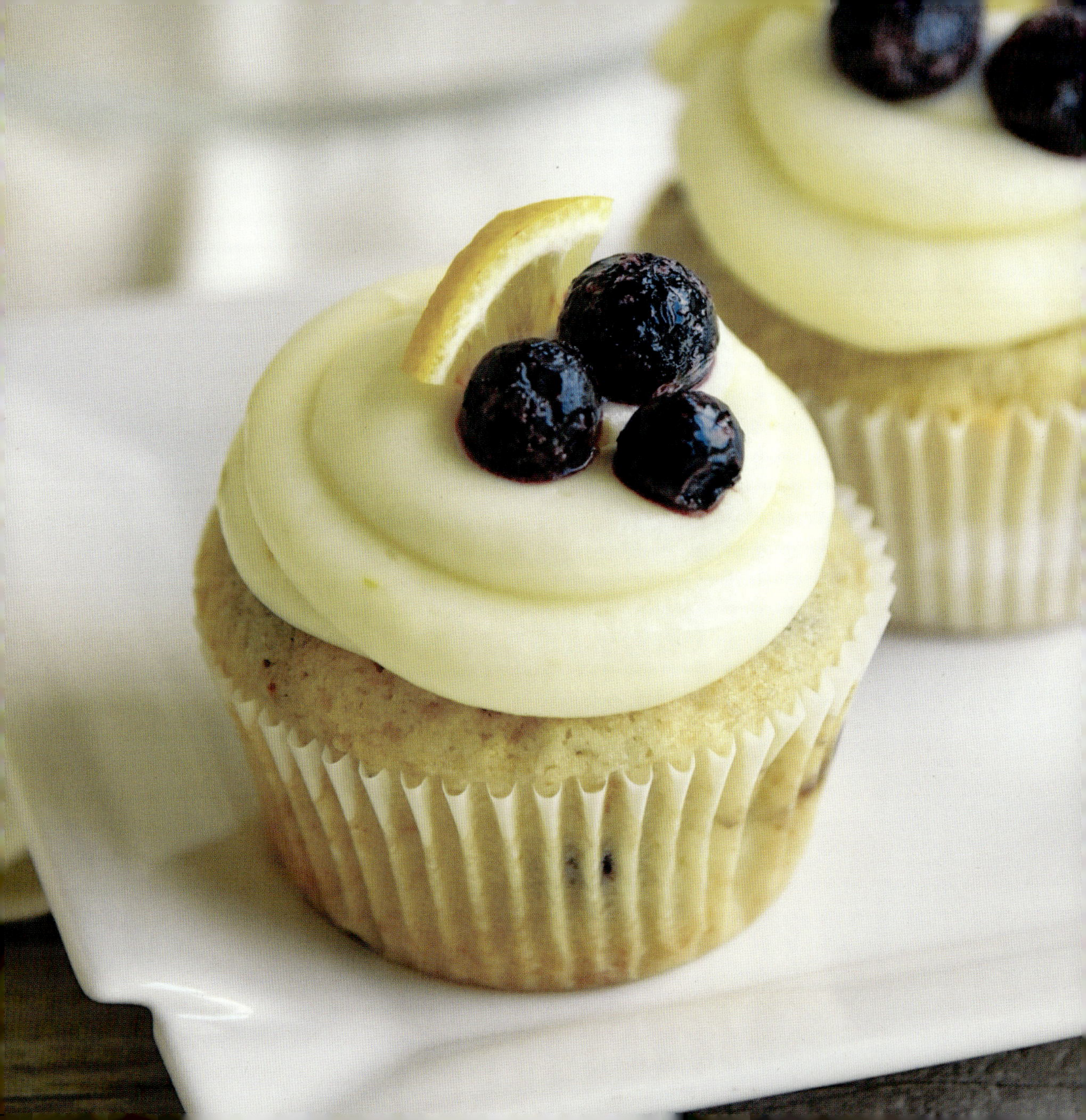

# Lemon-Blueberry Cupcakes

I like using wild blueberries in cupcakes because their smaller size distributes well without creating large pockets of cooked fruit in the small cakes. Because fresh wild blueberries have a short season and I don't live near a patch, I use frozen ones. I don't bother thawing them out before adding them to the cupcake batter, and it just adds a few extra minutes of baking time.

makes 1 dozen

### INGREDIENTS

½ cup butter, softened
1 cup granulated sugar
4 egg whites
2 teaspoons lemon zest
1 tablespoon fresh lemon juice
¾ cup fresh or frozen wild or small blueberries
1¾ cups plus 1 tablespoon all-purpose flour, divided
1½ teaspoons baking powder
½ teaspoon salt
½ cup milk
Lemon Cream Cheese Frosting (recipe at right)

### GARNISHES

fresh or frozen blueberries, lemon slices

Preheat oven to 325°. Place 12 muffin cup liners in a muffin pan.

Beat butter and sugar on high speed with an electric mixer until light and fluffy. Gradually beat in egg whites, zest, and juice.

Combine blueberries and 1 tablespoon flour in a small bowl, tossing to coat.

Combine remaining 1¾ cups flour, baking powder, and salt in a small bowl. Add flour mixture to butter mixture alternately with milk. Fold in blueberries. Spoon mixture into prepared muffin cups.

Bake 18 to 22 minutes or until center is dry when tested with a wooden pick (add 5 to 10 minutes if using frozen berries). Cool in pan 5 minutes; transfer to a wire rack and cool completely. Pipe or spread frosting over top. Garnish, if desired.

*Lemon Cream Cheese Frosting:* Beat **8 ounces softened cream cheese** and **½ cup butter** with an electric mixture until creamy. Gradually beat in **2 cups powdered sugar.** Gradually beat in **2 to 3 tablespoons fresh lemon juice,** scraping down sides of bowl if necessary.

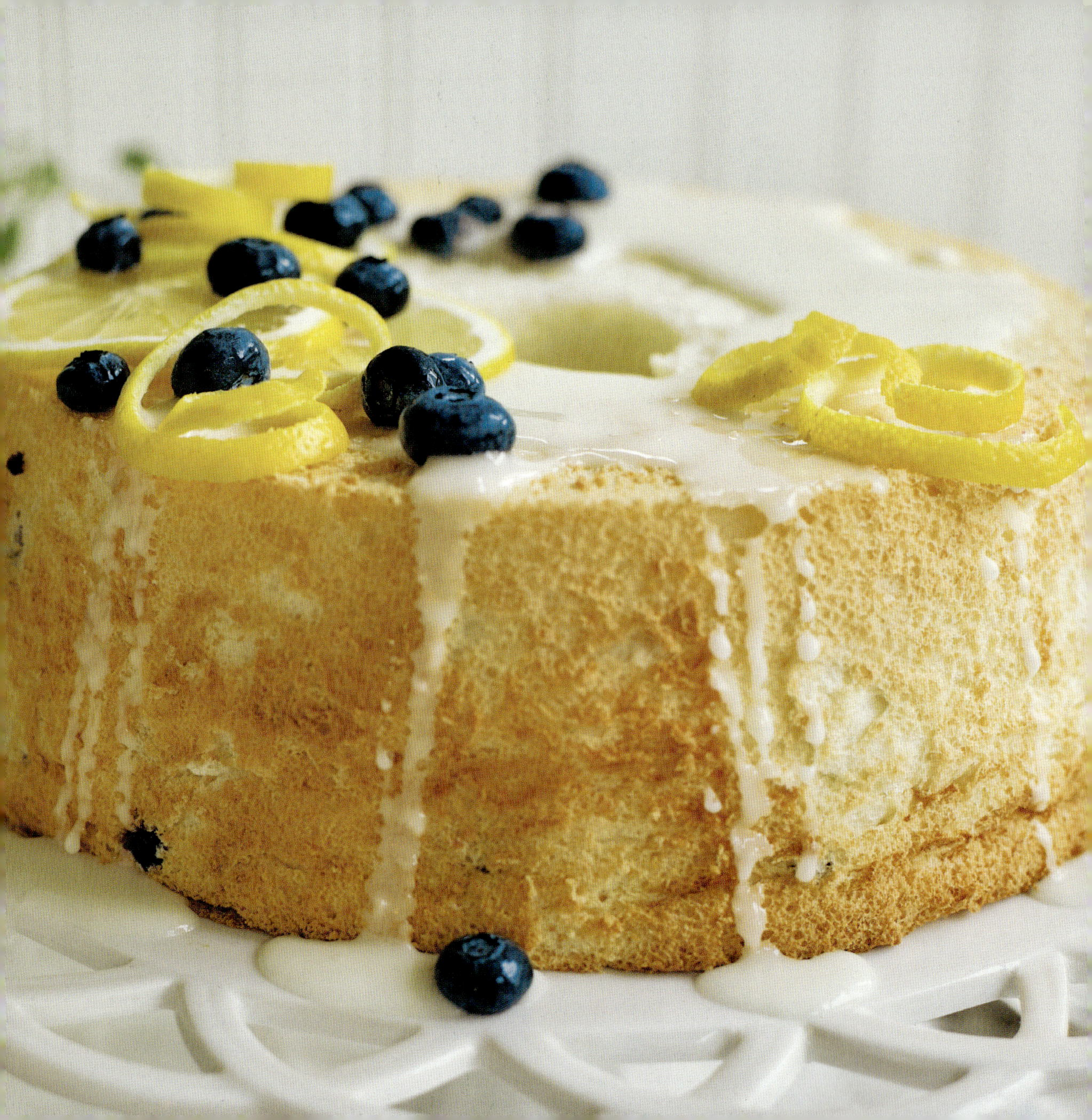

# Lemon-Blueberry Angel Food Cake

Cake flour is recommended for angel food cake, as it has a lower protein content and finer texture. You can get a pretty good-tasting cake with all-purpose flour, but a better substitution would be to replace 2 tablespoons of the flour with cornstarch.

makes 8–10 servings

### INGREDIENTS

1½ cups fresh or frozen blueberries
1 cup plus 1 tablespoon sifted cake flour, divided
12 large egg whites, at room temperature
¼ teaspoon salt
1 teaspoon vanilla extract
2 teaspoons fresh lemon juice
¾ teaspoon cream of tartar
1 cup granulated sugar
1 teaspoon lemon zest
1 cup sifted powdered sugar
3 tablespoons fresh lemon juice

### GARNISHES

lemon zest, fresh blueberries

Preheat oven to 350°.

Toss blueberries with 1 tablespoon flour in a medium bowl.

In a separate bowl, beat egg whites and salt at high speed with an electric mixer until foamy. Beat in vanilla and lemon juice. Add cream of tartar; beat until soft peaks form. Add granulated sugar, a few tablespoons at a time, beating until firm peaks form. Fold in remaining 1 cup flour, ¼ cup at a time. Fold in zest and blueberries.

Spoon batter into an ungreased 10-inch tube pan, spreading evenly. Cut through batter with a knife to break air pockets.

Bake 25 to 30 minutes until cake springs back when lightly touched. Invert pan; cool completely. Loosen cake from sides of pan, and invert onto a plate.

Combine powdered sugar and lemon juice in a small bowl, whisking until smooth. Drizzle over cooled cake and garnish, if desired.

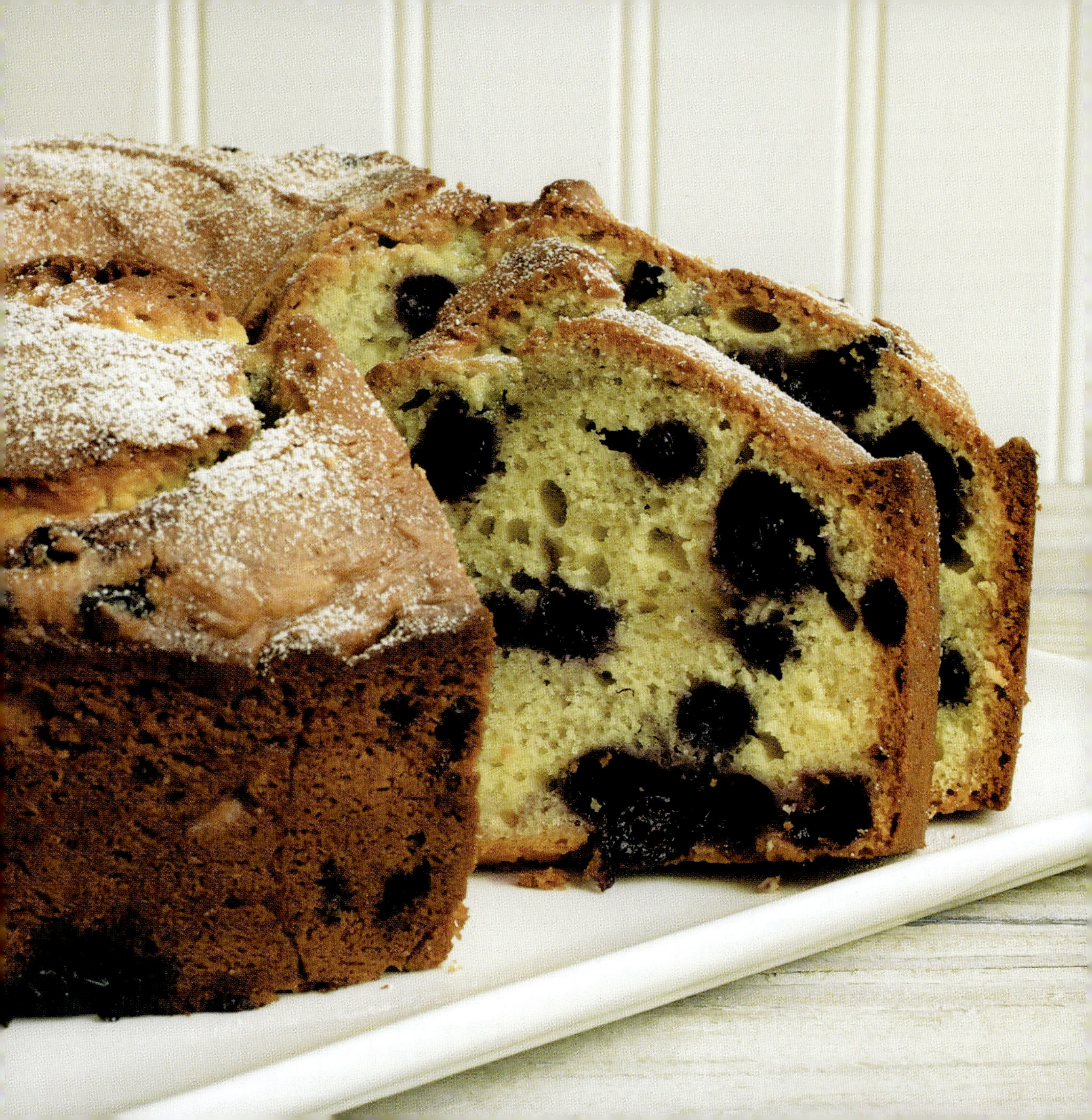

# Blueberry Sour Cream Pound Cake

makes 8–12 servings

INGREDIENTS

2 cups blueberries
3 cups all-purpose flour, divided
1 teaspoon baking powder
½ teaspoon baking soda
½ teaspoon salt
½ cup butter
½ (8-ounce) package cream cheese, softened
2 cups granulated sugar
3 large eggs
1 (8-ounce) container sour cream
2 teaspoons vanilla extract

Preheat oven to 350°. Line a 9x5-inch loaf pan with nonstick aluminum foil or lightly greased foil.

Combine blueberries and 2 tablespoons flour in a small bowl, tossing to coat.

In a medium-size bowl, combine remaining flour, baking powder, baking soda, and salt.

Beat butter, cream cheese, and sugar in a large bowl with an electric mixer. Beat in eggs, one at a time. Beat in sour cream and vanilla. Add flour mixture, and fold in blueberries.

Spoon batter into prepared pan. Bake 1 hour–1 hour and 10 minutes or until center is dry when tested with a wooden pick. Cool in pan 10 minutes. Remove from pan and cool completely on a wire rack.

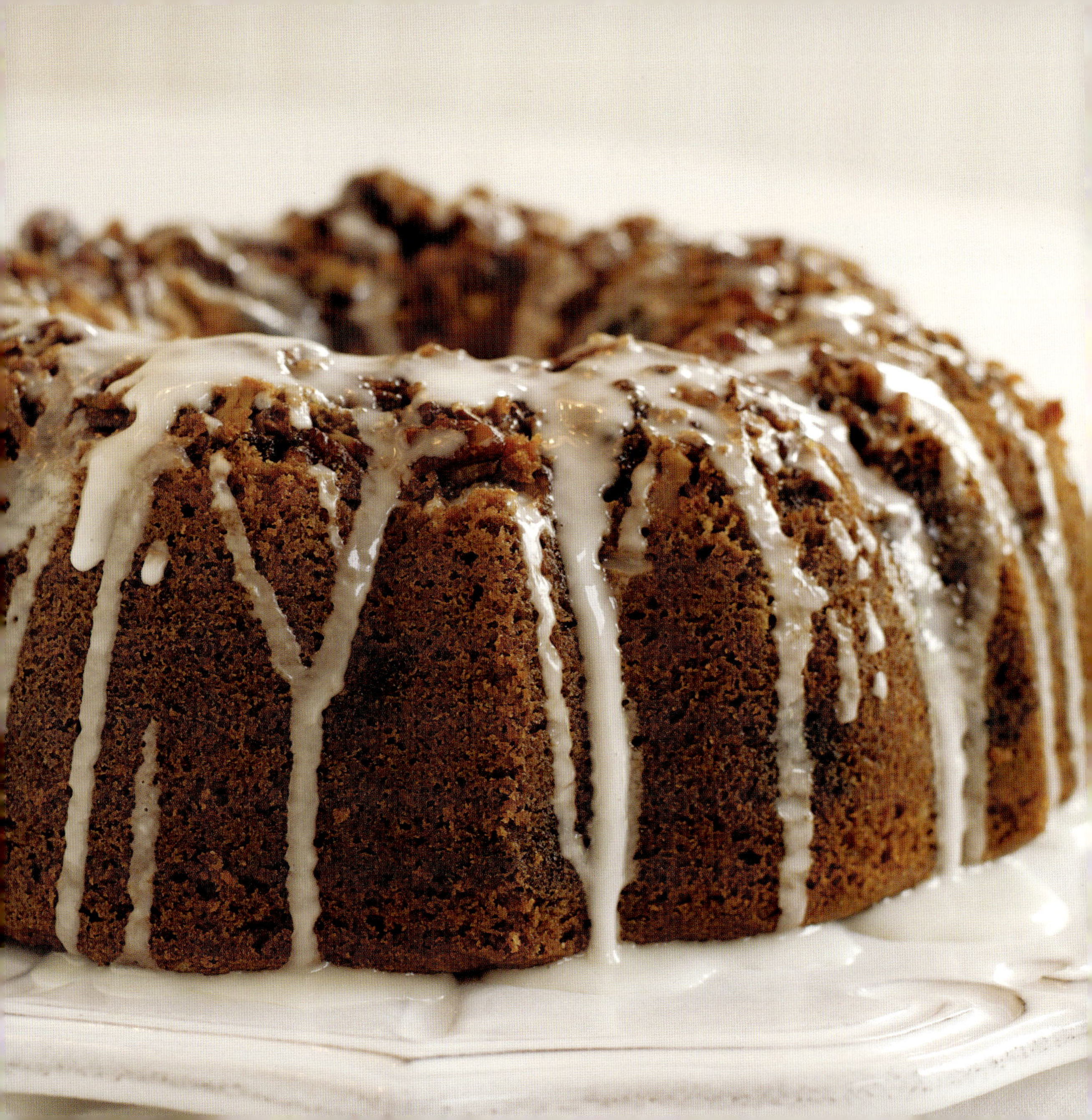

# Blueberry Sour Cream Coffee Cake

makes 1 (10-inch) cake

### INGREDIENTS

1 cup pecans
3 tablespoons firmly packed light brown sugar
3 tablespoons melted butter
2 cups fresh or frozen (and thawed) blueberries
2½ cups plus 1 tablespoon all-purpose flour, divided
1 cup butter, softened
1½ cups granulated sugar
1½ teaspoons lemon zest
3 large eggs
1 teaspoon vanilla extract
1 tablespoon baking powder
½ teaspoon baking soda
¾ teaspoon salt
1½ cups sour cream
1½ cups powdered sugar
2 tablespoons heavy cream or milk

Preheat oven to 350°. Spread pecans on a baking sheet in a single layer. Bake 5 to 8 minutes or until toasted and fragrant. Cool slightly and coarsely chop. Combine pecans, brown sugar, and melted butter in a small bowl. Spread pecan mixture on bottom of well-buttered 10-inch Bundt pan. Toss blueberries in 1 tablespoon flour in a small bowl; set aside.

Cream softened butter and granulated sugar in a large mixing bowl with an electric mixer on medium-high speed. Beat in zest. Beat in eggs, one at a time. Beat in vanilla.

In a separate bowl, combine remaining 2½ cups flour, baking powder, baking soda, and salt. Beat flour mixture into butter mixture, alternating with sour cream.

Spoon about one-third batter over pecan topping mixture. Top with about ¾ cup blueberries. Alternate batter and blueberries, ending with batter mixture.

Bake 1 hour to 1 hour and 10 minutes or until a wooden pick inserted in center comes out clean. Cool 15 minutes.

Run a knife around edges of cake. Place a cake plate over pan and invert. Whisk together powdered sugar and cream in a small bowl. Drizzle over cake.

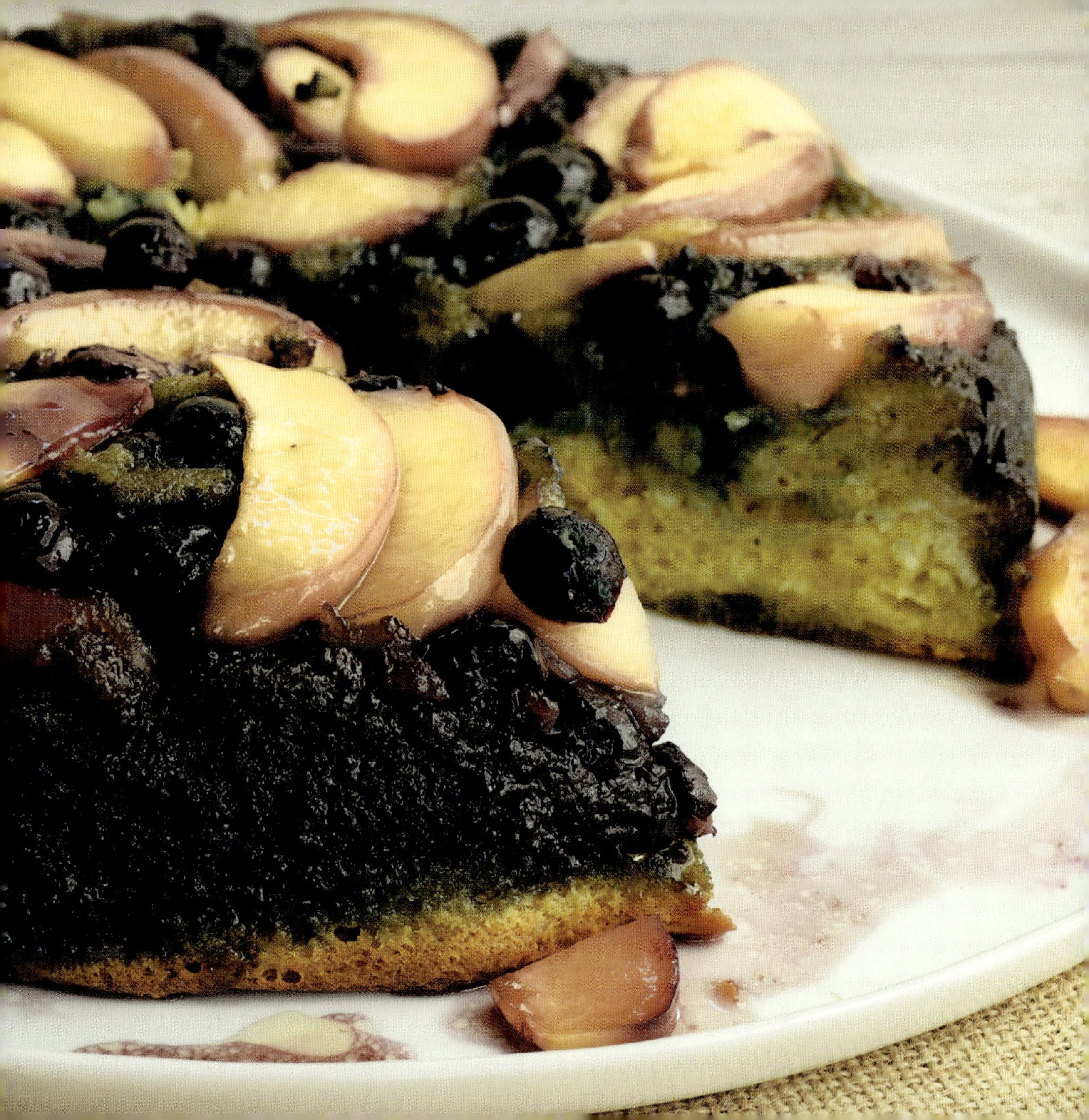

# Blueberry-Nectarine Upside-down Cake

makes 8 servings

INGREDIENTS

¾ cup butter, divided
¾ cup firmly packed light brown sugar
2 cups fresh nectarines or peeled peaches, sliced
1 cup fresh or frozen (and thawed) blueberries
2 cups all-purpose flour
1 teaspoon baking soda
¼ teaspoon salt
1 cup buttermilk
2 large eggs
1 teaspoon vanilla extract
½ cup granulated sugar

Preheat oven to 350°.

Melt 6 tablespoons butter in a 10-inch cast iron or heavy ovenproof skillet over medium heat. Tilt skillet to coat sides. Remove from heat and stir in brown sugar. Arrange nectarine slices and blueberries on bottom of skillet.

Stir together flour, baking soda, and salt in a large bowl.

Melt remaining 6 tablespoons butter. Combine melted butter, buttermilk, eggs, vanilla, and granulated sugar, whisking vigorously. Gradually add egg mixture to flour mixture, stirring until well blended. Spread batter evenly over fruit.

Bake 40 minutes or until a wooden pick inserted in center comes out almost clean. Cool in pan 10 minutes. Invert cake onto a round platter. Scrape any sugar mixture left in the skillet over the cake. Serve warm or at room temperature.

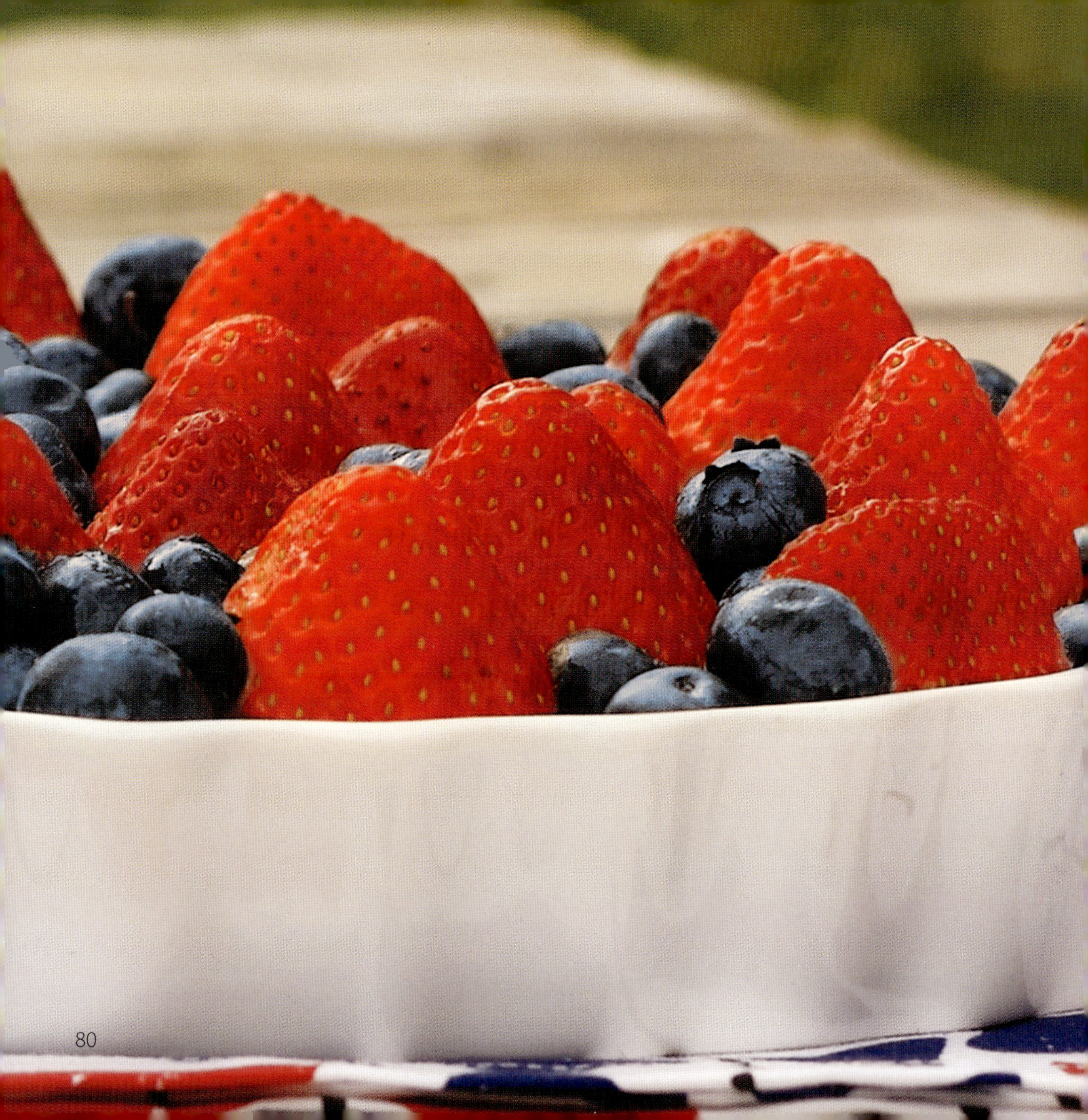

# pies and tarts

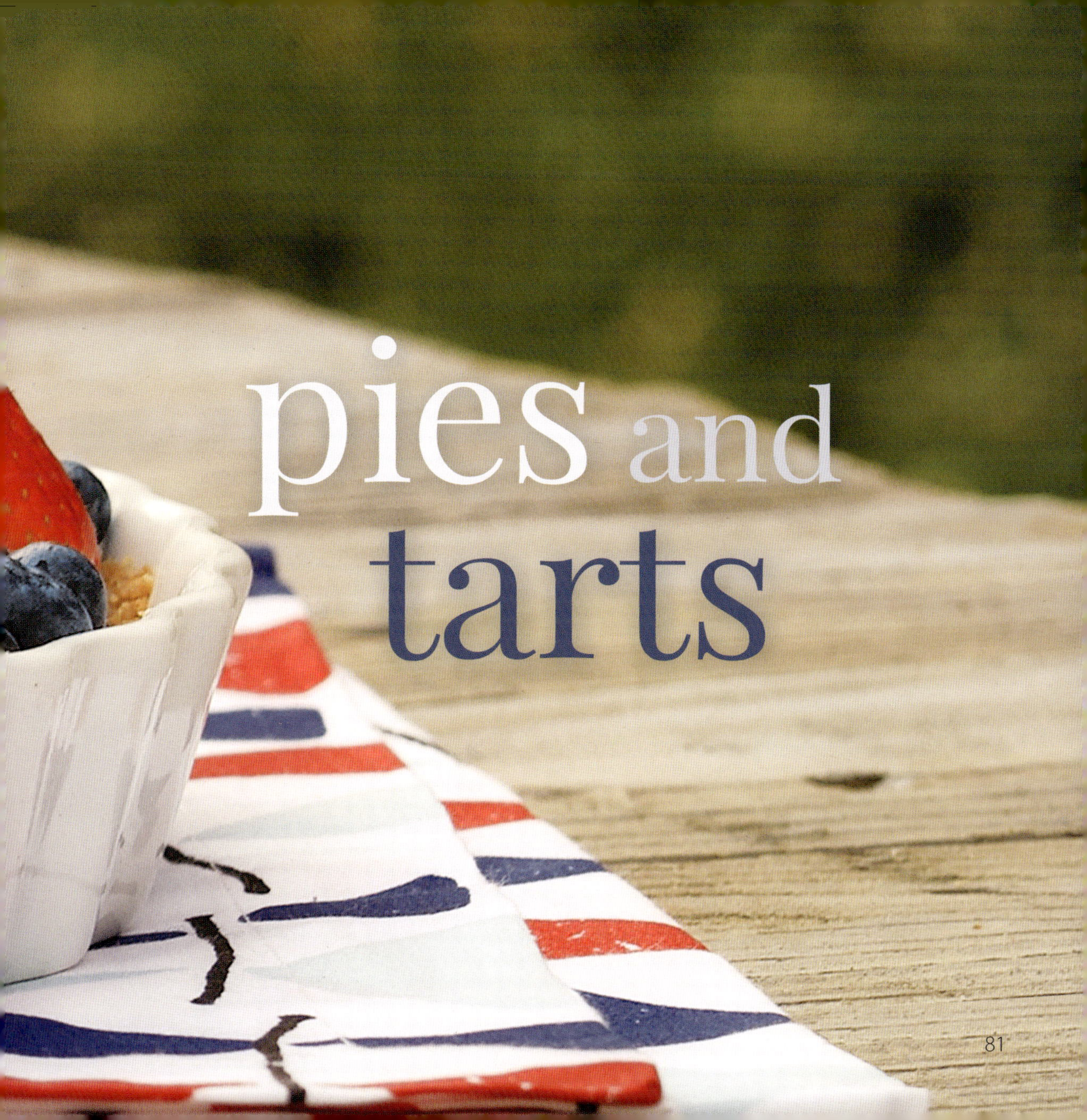

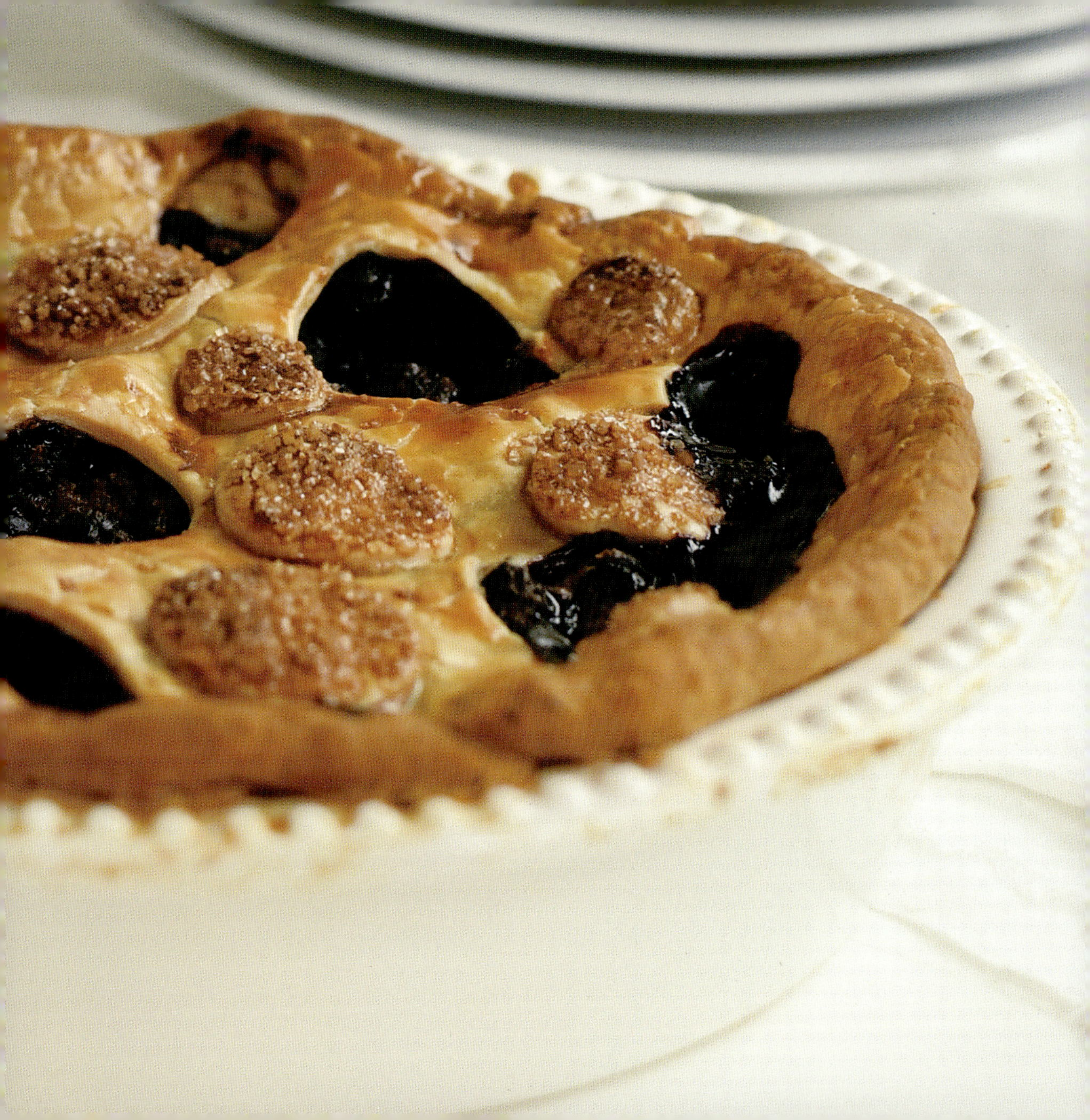

# Spiced Blueberry Pie

Be sure to make the crust early—it needs to chill. The crust can be made ahead and stored in the refrigerator up to 3 days or in the freezer for about 3 months.

makes 8 servings

INGREDIENTS

1 cup granulated sugar
¼ cup tapioca starch or instant tapioca
1 teaspoon lemon zest
1 tablespoon fresh lemon juice
1½ teaspoons ground cinnamon
¾ teaspoon ground nutmeg
¼ teaspoon ground cloves
4 cups fresh blueberries
Double-Crust Pastry Dough (recipe at right) or 1 package store-bought refrigerated piecrusts
Butter
1 large egg
1 tablespoon water
1 tablespoon turbinado sugar (optional)

Stir together granulated sugar, tapioca, zest, juice, cinnamon, nutmeg, and cloves in a large bowl. Add blueberries, tossing until well blended. Let filling stand 15 minutes.

Preheat oven to 400°.

On a lightly floured surface, roll 1 dough disk into a 12-inch circle. Fit dough into a buttered 9-inch pie plate. Spoon filling into piecrust. Roll remaining dough disk into an 11-inch circle. Place over filling. Press edges of dough together, fold under, and pinch to seal. Cut several slits or shapes in top.

Combine egg and water in a small bowl; brush egg mixture over pastry top and sprinkle with turbinado sugar, if desired. Bake 45 to 50 minutes or until thickened and bubbly. Cool on a wire rack, and serve warm or cool.

*Double-Crust Pastry Dough:* Combine **3 cups all-purpose flour, 2 tablespoons sugar,** and **½ teaspoon salt** in a food processor, pulsing until blended. Add **1 cup cold butter,** cut into pieces; pulse until mixture resembles coarse sand. Combine **½ cup cold water** and **2 teaspoons vinegar.** Sprinkle half of water mixture over flour mixture; pulse to combine. Sprinkle in remaining water mixture; pulse until mixture comes together. Transfer to a lightly floured surface; press together until blended. Roll into 2 disks and wrap with plastic wrap. Chill 1 hour.

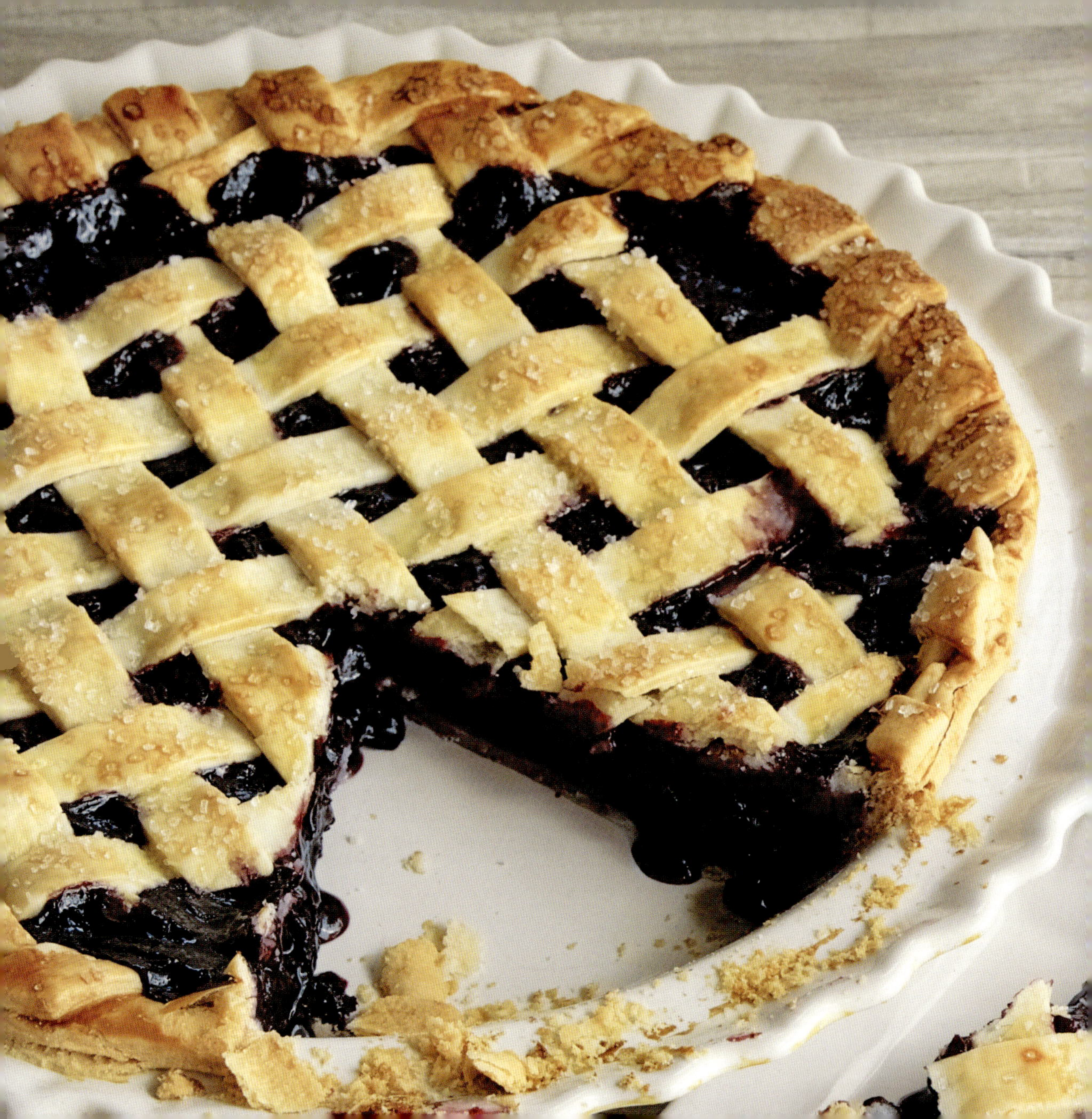

# Double-Crust Lattice Blueberry Pie

Using frozen berries in pies produces erratic results. Often, they weep too much as they cook, even if they are thawed and drained before using in a pie filling. The flavor with frozen berries is good, but the pie doesn't slice cleanly and the filling is too runny. To get the firmer texture and ease of slicing I prefer, I cook the filling prior to putting it in the oven. This recipe is a bit of a hybrid—calling for less expensive frozen berries in the cooked portion of the filling but stirring in fresh berries at the end. The fresh berries soften as the pie bakes, but they don't fall apart.

makes 1 (9-inch) pie or 8 servings

INGREDIENTS

4 cups fresh blueberries, divided
¾ cup granulated sugar
1 teaspoon lemon zest
2 tablespoons fresh lemon juice
3 tablespoons cornstarch
2 tablespoons water
1 tablespoon butter
Double-Crust Pastry Dough (page 83) or 1 package store-bought refrigerated piecrusts
1 egg, lightly beaten
Sparkling or turbinado sugar (optional)

Combine 2 cups blueberries (frozen are okay here), granulated sugar, zest, and juice in a small saucepan. Cook over medium heat 8 to 10 minutes, pressing berries against pan with the back of a spoon to crush. Combine cornstarch with 2 tablespoons water in a small bowl; stir into berry mixture. Bring to a boil; reduce heat and simmer 2 minutes or until thickened. Remove from heat and add butter, stirring until well blended. Stir in remaining 2 cups fresh blueberries. Set filling aside.

Preheat oven to 375°.

On a lightly floured surface, roll 1 dough disk into an 11-inch circle. Fit into a 9-inch pie plate. Roll remaining dough disk into a 9-inch circle. Cut into ¾-inch strips.

Spoon filling into crust. Weave strips into a lattice pattern over top. Brush crust dough with egg and sprinkle with sparkling sugar, if desired. Bake 40 minutes or until golden brown and bubbly, shielding crust with aluminum foil after 20 minutes.

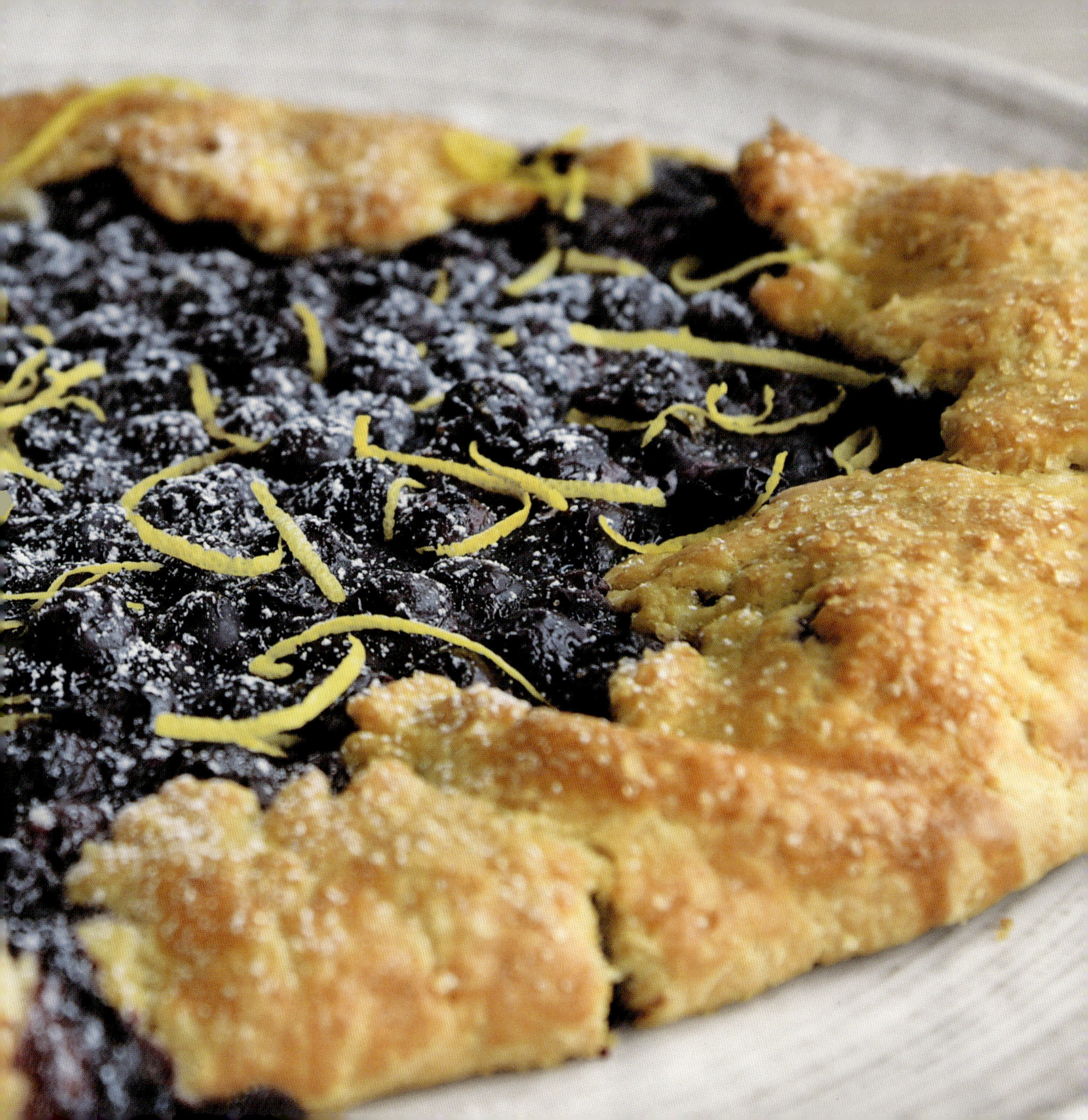

# Blueberry Free-form Pie

For pie lovers who want a lot of crust, this one's for you! Buy the berries in bulk for this treat—a great place to look are the big-box clubs. Be sure to make the crust early—it needs to chill. You can make the crust ahead and refrigerate it for up to 3 days or freeze it for about 3 months.

makes 6–8 servings

### INGREDIENTS

Double-Crust Pastry Dough (page 83)
½ cup granulated sugar
¼ cup tapioca
1 teaspoon lemon zest
1 tablespoon fresh lemon juice
4 cups fresh blueberries
1 egg, lightly beaten
Sparkling or turbinado sugar (optional)

### GARNISHES

powdered sugar, lemon zest

Preheat oven to 375°.

Place dough disk on counter for 5 to 10 minutes to soften slightly. (Don't let it get warm.) Line a rimmed baking sheet with nonstick aluminum foil, parchment paper, or a silicone baking mat.

Combine granulated sugar, tapioca, lemon zest, and juice in a large bowl. Stir in berries. Let stand 5 minutes, stirring occasionally to moisten dry ingredients.

On a lightly floured surface, roll dough disk to form a long rectangle (a little more than ⅛-inch thick). Transfer to prepared pan.

Spoon berry mixture into center of piecrust, leaving several inches of crust around edges. Fold crust over edges of pie, leaving center of pie exposed. Brush egg over outer edges of crust and sprinkle with sparkling sugar, if desired.

Bake 45 to 50 minutes or until crust is golden brown and center is bubbling. Garnish, if desired. Serve warm or cool.

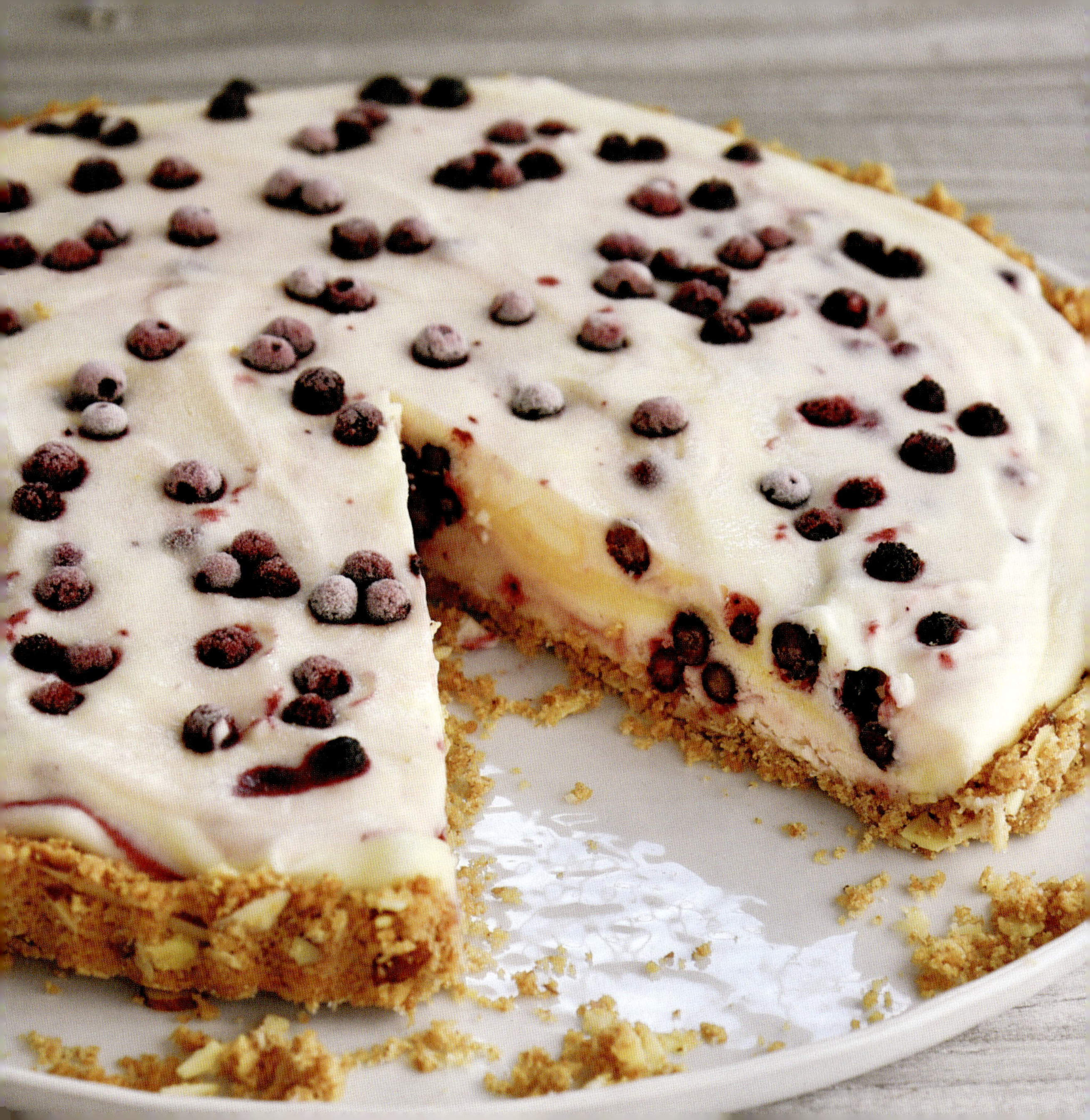

# Frozen Blueberry-Key Lime Pie

This pie is great for parties when you end up spending a lot of energy on the dinner and need a no-fuss but impressive finale. Keep it frozen until a few minutes before serving—just enough time to clear the dishes off the table. For easiest slicing, use a tart pan with a removable bottom. The creamy-tart flavor of the pie is just as delicious prepared in a traditional pie plate, but that first slice can get a bit messed up when removed. I like using the small wild blueberries here because they are usually found frozen.

makes 8 servings

INGREDIENTS

Almond-Graham Crust (recipe at right)
1½ cups granulated sugar
½ cup butter, cut into pieces
1 cup Key or Persian lime juice
¼ teaspoon salt
4 large eggs
1 cup heavy whipping cream
1½ cups fresh or frozen, wild or purchased blueberries, divided

Prepare crust; set aside. Combine sugar, butter, juice, and salt in a large heavy saucepan over medium heat. Cook, stirring constantly, 3 minutes or until sugar dissolves and butter melts.

Whisk eggs in a medium-size bowl. Whisk ¼ of the hot juice mixture into eggs; add egg mixture back into remaining juice mixture. Cook over medium-low heat, whisking constantly, 5 minutes or until mixture reaches 160° and coats the back of a spoon. Remove from heat. Cover with plastic wrap; chill.

Whip cream in a large bowl on high speed until soft peaks form. Spoon a dollop into lime mixture, stirring until mixture lightens in texture. Gently fold lime mixture into remaining whipped cream. Fold in berries, reserving some for top. Spoon berry mixture into crust and top with reserved berries; freeze several hours or until firm.

*Almond-Graham Crust:* Combine **1½ cups graham cracker crumbs, ½ cup chopped sliced almonds,** and **¼ cup sugar** in a large bowl; stir in **6 tablespoons melted butter.** Press into a 10-inch removable-bottom tart pan. Bake at 350° for 10 minutes.

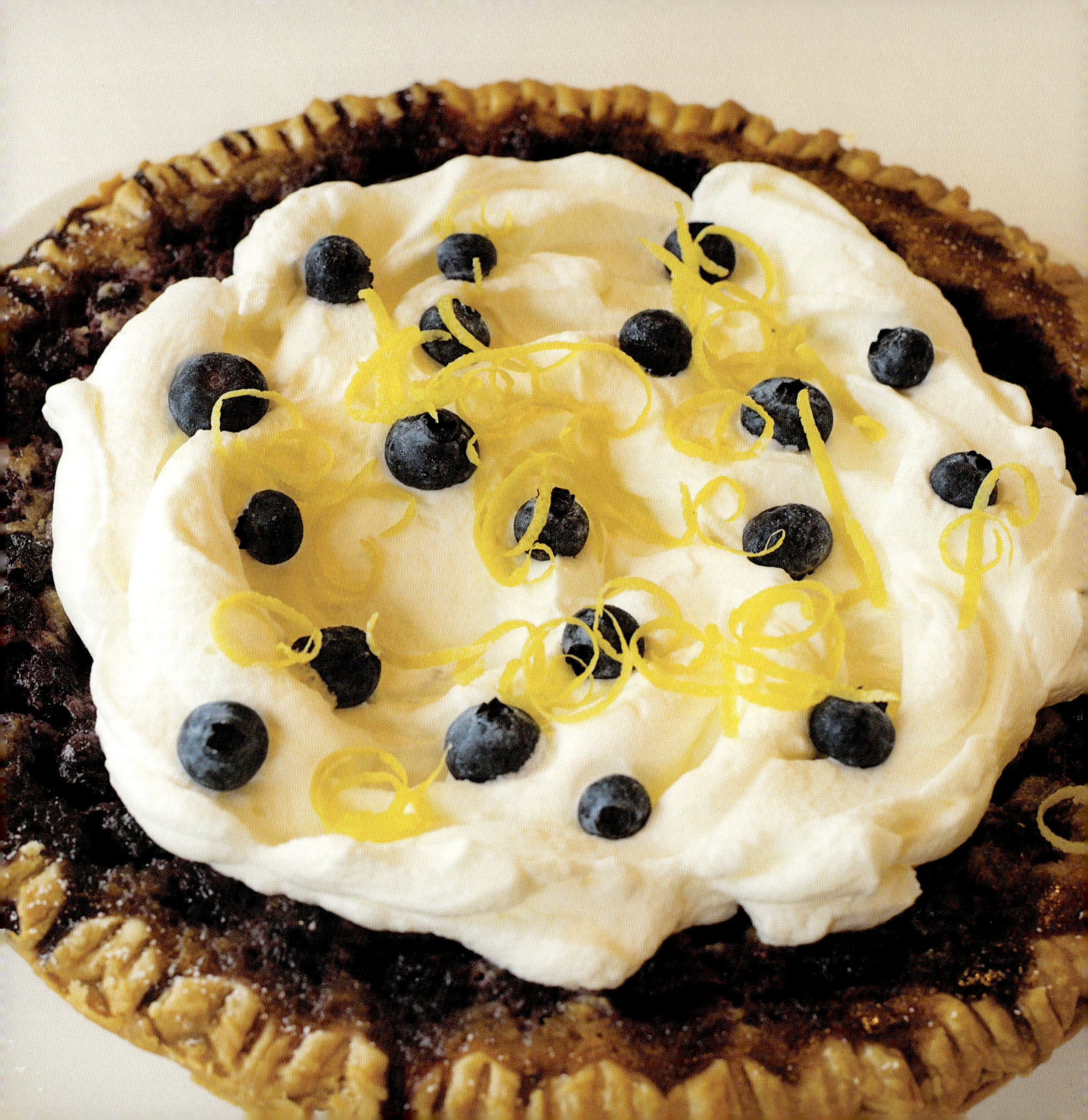

# Blueberry-Buttermilk Pie

makes 1 (9-inch) pie

INGREDIENTS

3 large eggs
3 tablespoons all-purpose flour
1 cup granulated sugar
1 cup buttermilk
6 tablespoons butter, melted
2 teaspoons lemon zest
2 tablespoons fresh lemon juice
2 cups fresh or frozen blueberries
Basic Piecrust Dough (recipe at right) or 1 refrigerated or frozen deep-dish piecrust, thawed
Sweetened whipped cream (optional)

GARNISHES

lemon curls, fresh blueberries

Preheat oven to 375°.

Whisk eggs in a large bowl. Whisk in flour, sugar, buttermilk, butter, lemon zest, and lemon juice.

Spread blueberries in bottom of prepared crust. Pour egg mixture over berries.

Bake 50 minutes, shielding edges of pie with aluminum foil, if necessary, to prevent overbrowning. Pie will be a little soft (but not liquid) in center and will firm up as it cools. Top with whipped cream, and garnish, if desired.

*Basic Piecrust Dough:* Combine **1½ cups all-purpose flour** and **¼ teaspoon salt** in a food processor; pulse until blended. Add **¾ cup cold butter, cut into pieces,** and pulse several times until crumbly. With processor running, drizzle in **2 to 4 tablespoons ice water,** 1 tablespoon at a time. Process just until dough forms a ball and pulls away from sides of bowl. If necessary, roll into a disk and chill until firm yet pliable. Piecrust may be made ahead, covered tightly with plastic wrap, and refrigerated up to a week or frozen up to 3 months. Roll to ⅛-inch thickness on a lightly floured surface and place in pie plate.

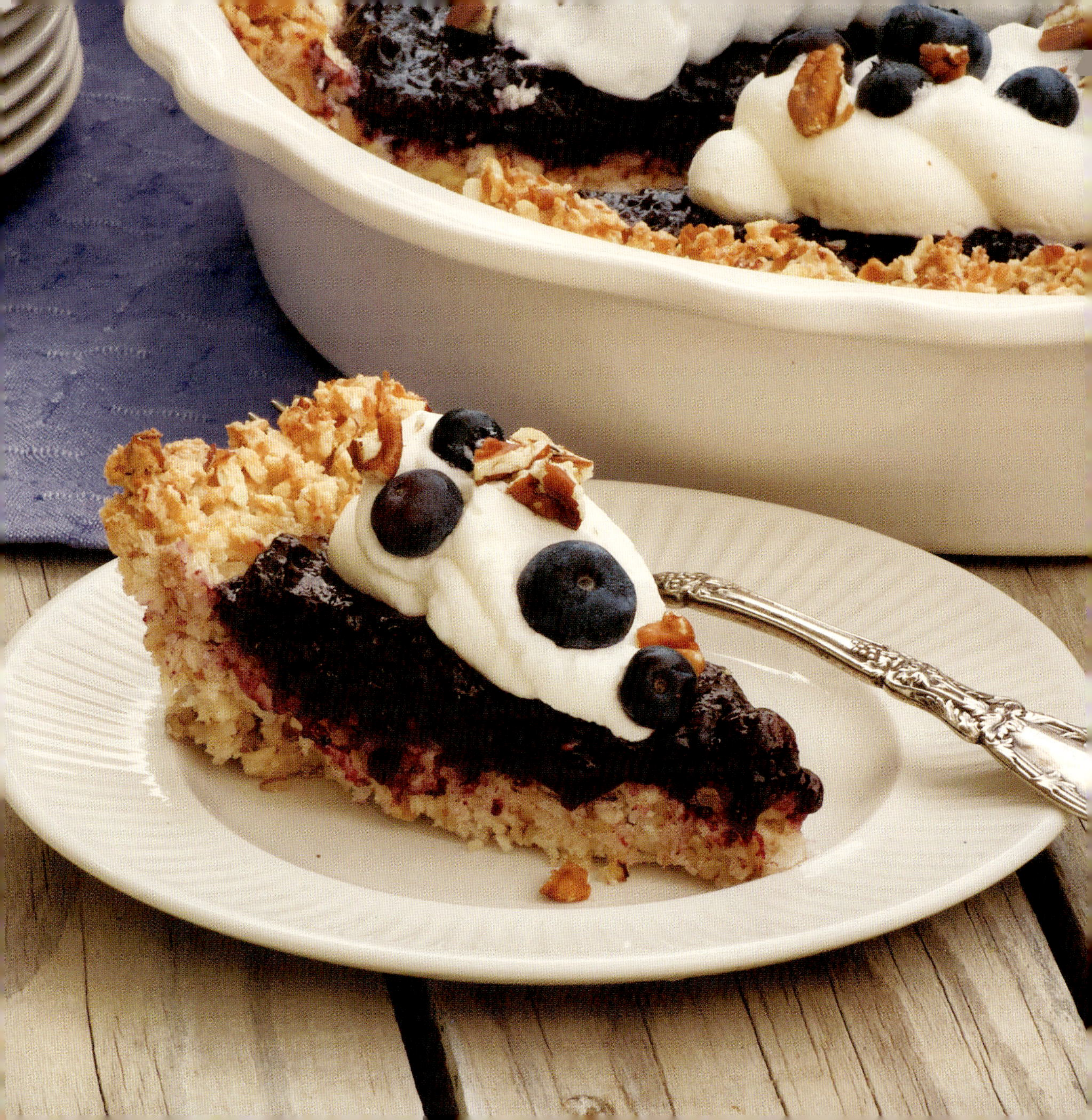

# Southern Comfort Pie

Gwen York's blueberry dessert is one you'll definitely want to keep in short rotation. With the salty-sweet crust and sturdy filling, you can cut it into slender slices and serve this pie at potlucks and buffets with no worries about the dessert falling apart. Just serve the whipped cream on the side. Gwen is Lynne Ryan's mom (her recipes are on pages 127 and 129). It' s clear that good food and Southern hospitality run in this family!

makes 1 (9-inch) pie

## INGREDIENTS

Coconut-Pecan Crust (recipe at right)
½ cup granulated sugar
2 tablespoons cornstarch
¼ teaspoon salt
½ cup water
2 cups fresh or frozen blueberries
1 teaspoon lemon zest
1 tablespoon fresh lemon juice
2 cups heavy whipping cream
¼ cup powdered sugar
1 teaspoon vanilla extract

## GARNISHES

chopped pecans, fresh blueberries

Prepare Coconut-Pecan Crust; cool completely.

Combine sugar, cornstarch, and salt in a medium-size saucepan. Gradually whisk in ½ cup water until smooth. Stir in blueberries, zest, and juice. Bring to a boil over medium heat, stirring constantly. Cook 2 to 3 minutes, stirring occasionally, until thickened and bubbly (some berries will burst, many will remain whole). Remove from heat and cool completely.

Spoon blueberry filling into prepared crust. Cover and chill until ready to serve. Before serving, combine cream, powdered sugar, and vanilla in a large bowl. Beat together or whisk until soft peaks form. Top pie with whipped cream or serve on the side. Garnish, if desired.

*Coconut-Pecan Crust:* Combine **3 egg whites** (at room temperature) and **¼ teaspoon cream of tartar** in a large bowl. Beat or whisk until stiff peaks form. Fold in **1 cup sugar, 1¾ cups saltine cracker crumbs** (1 sleeve, crushed), **1 cup sweetened flaked coconut,** and **1 cup chopped pecans.** Press mixture in bottom and up sides of a buttered 9-inch-deep pie plate. Bake at 350° for 25 minutes or until golden brown and firm. Cool completely.

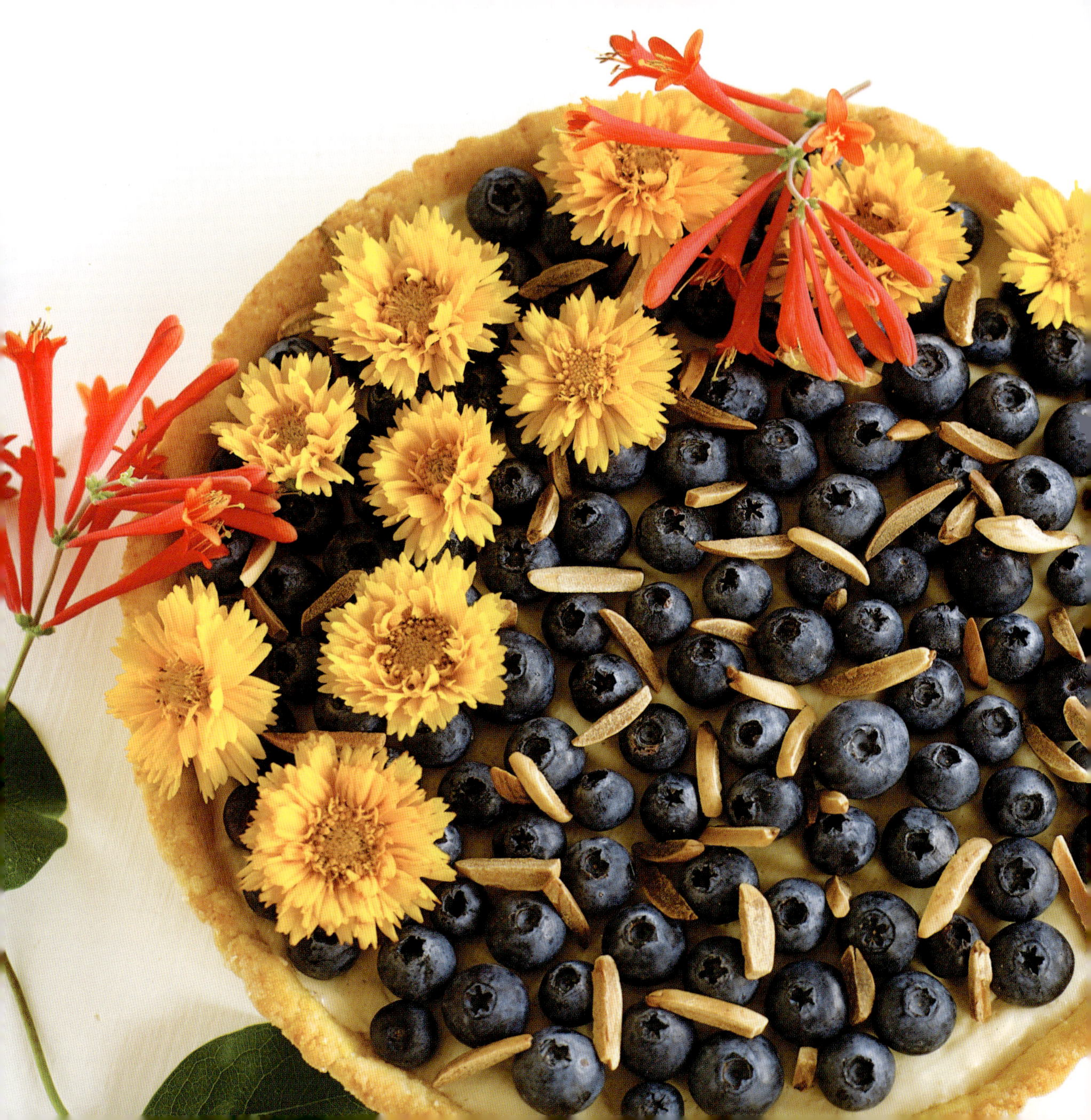

# Keto-Friendly Berry Almond Cream Tart

Almond flour, almond meal, and coconut flour can be great substitutes for all-purpose flour. I like to add a bit of coconut flour to the mix, but, if it's unavailable, you can substitute almond meal (which has more texture than almond flour) or use all almond flour. If eating gluten-free isn't necessary or appealing to you, simply use a traditional piecrust.

makes 1 (9-inch) pie

### INGREDIENTS

- Basic Gluten-Free Piecrust Dough (recipe at right)
- 1 (8-inch) package cream cheese, softened
- ½ cup firmly packed light brown sugar substitute or brown sugar
- ⅓ cup sour cream
- ¼ cup heavy whipping cream
- 2 tablespoons fresh lemon juice
- 1 teaspoon vanilla extract
- 1 cup fresh blueberries

### GARNISH

- fresh edible flowers, slivered almonds

Preheat oven to 375°. Roll out piecrust dough and place in a 9-inch tart pan. Prick bottom and sides of crust with a fork. Bake 8 to 10 minutes or until golden brown. Cool completely.

Beat cream cheese, brown sugar, and sour cream in a large mixing bowl until smooth and well blended. Beat in whipping cream, lemon juice, and vanilla.

Spread mixture in cooled piecrust with an offset spatula until smooth on top. Arrange berries on top. Garnish, if desired, and chill until filling is firm.

*Basic Gluten-Free Piecrust Dough:* Combine **1⅓ cups almond flour; ⅔ cup coconut flour** (or almond meal); **2 tablespoons powdered sugar substitute** (or powdered sugar); **1 large egg; ¼ cup butter, cut into pieces** (or coconut oil); and **⅛ teaspoon salt** in a mixing bowl. Beat at medium speed until mixture is crumbly. Drizzle in **1 to 3 teaspoons cold water,** beating at slow speed until mixture forms large clumps and holds together when squeezed. Pat dough into a thin disk. Cover and chill at least 30 minutes or until firm.

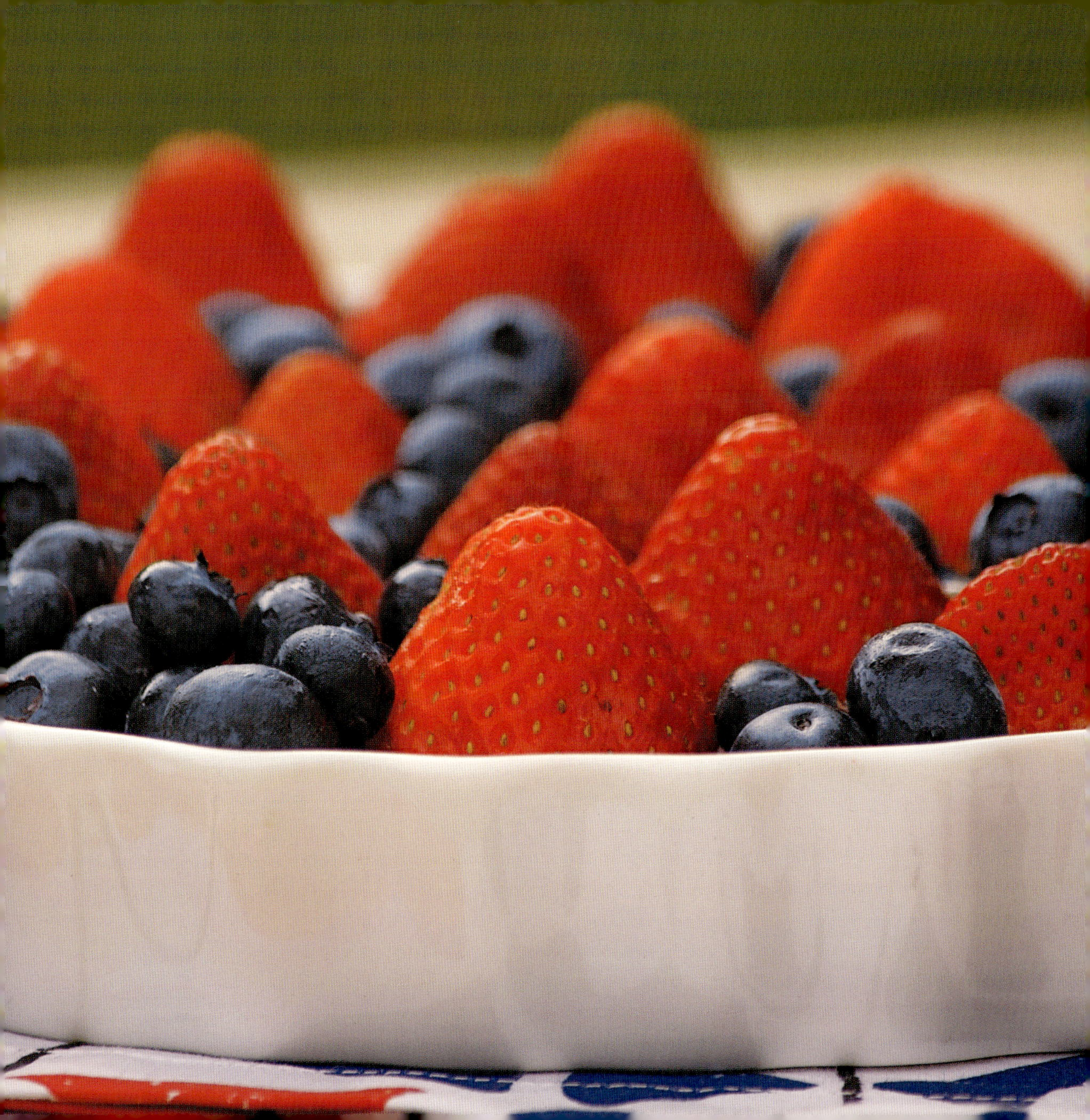

# Berry-Berry Chocolate Truffle Tart

makes 1 (9-inch) tart

CRUST
¾ cup graham cracker crumbs
¾ cup crushed shortbread cookies
¼ cup granulated sugar
⅓ cup butter, melted

FILLING
1¼ cups heavy whipping cream
12 ounces sweet dark or semi-sweet chocolate, chopped
2 teaspoons vanilla extract

TOPPINGS
1 (16-ounce) container fresh strawberries, hulled
1½ cups fresh blueberries

Preheat oven to 375°.

To make crust, combine cracker and cookie crumbs in a medium bowl. Add sugar and butter, stirring until well blended. Press into an ungreased 9-inch tart pan.

Bake 8 to 10 minutes; cool completely on a wire rack.

To make filling, heat cream in a medium saucepan over medium heat until hot but not boiling. Remove from heat. Add chocolate, stirring until smooth. Stir in vanilla. Pour into crust and set aside to cool.

When filling is cool, but not completely set, arrange strawberries evenly on top. Sprinkle with blueberries. Refrigerate until chilled and firm.

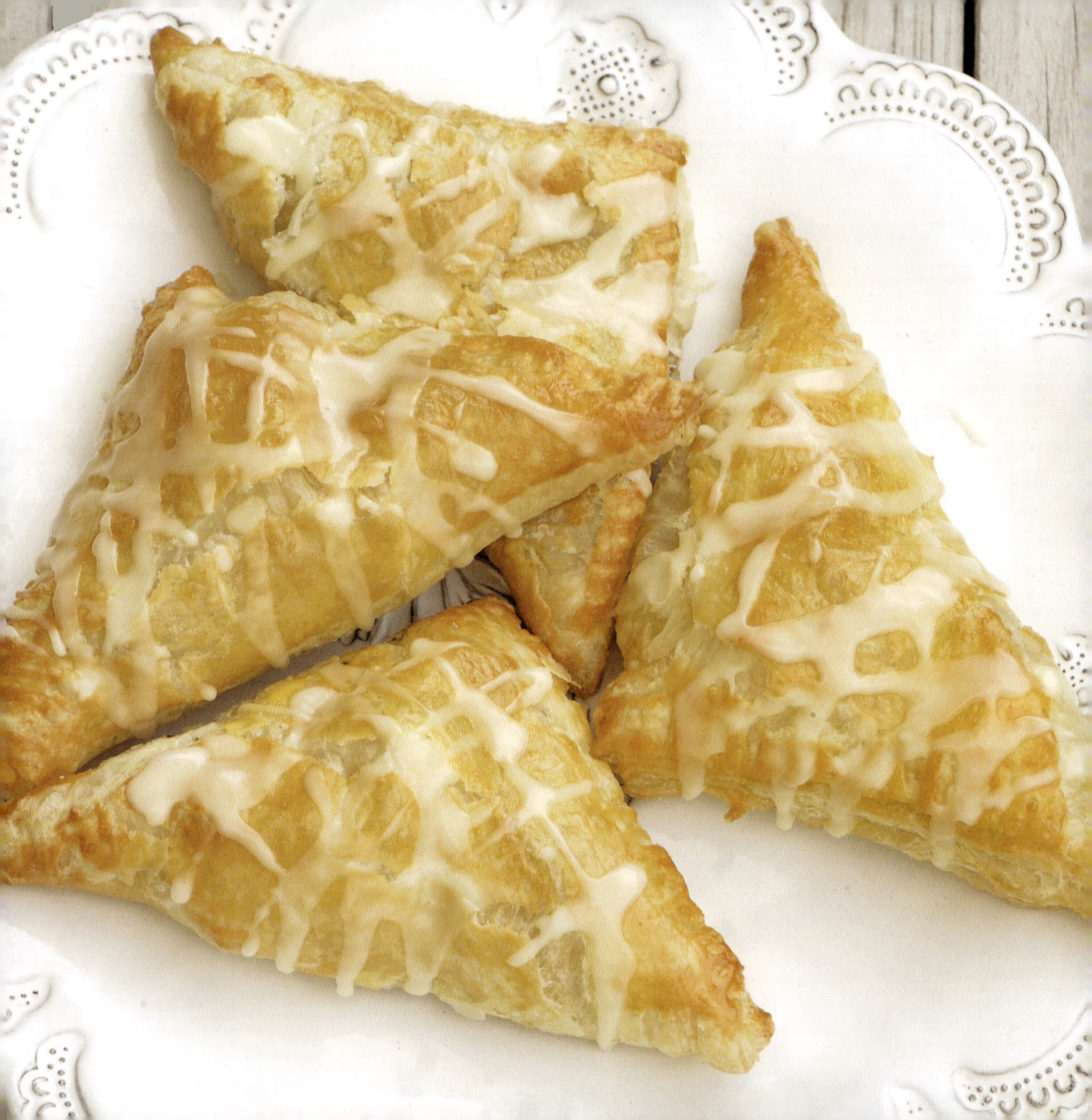

# Blueberry-Orange Cream Cheese Turnovers

I love these super-easy turnovers warm and fresh out of the oven. Keep in mind that the tangy-sweet frosting clings best to the top of cooled pastries, while it thins and rolls right off of warm ones. Once cooled, you can freeze turnovers; reheat them in the microwave or toaster oven when you're ready.

makes 4 servings

INGREDIENTS

4 ounces cream cheese, softened
2 tablespoons granulated sugar
½ teaspoon organic orange zest
½ teaspoon vanilla extract
½ (17.3-ounce) package frozen puff pastry sheets, thawed
¼ cup fresh or frozen blueberries
1 large egg, separated
⅓ cup powdered sugar
2 teaspoons orange juice

Preheat oven to 375°.

Combine cream cheese, granulated sugar, zest, and vanilla in a small bowl; set aside.

Roll out pastry sheets on a lightly floured surface. Cut each sheet into 4 squares. Spoon about 2 tablespoons cream cheese mixture in center of each pastry piece. Top with 1 tablespoon blueberries.

Brush edges with egg yolk, fold into triangles, and press to seal. Place on a parchment- or nonstick aluminum foil-lined baking sheet.

Bake pastries 15 to 20 minutes or until golden brown and puffed. Cool on a wire rack. Whisk together powdered sugar and orange juice in a small bowl. Drizzle over pastries.

*Variation:* Add about 1 teaspoon chopped dark chocolate or chocolate chips to filling.

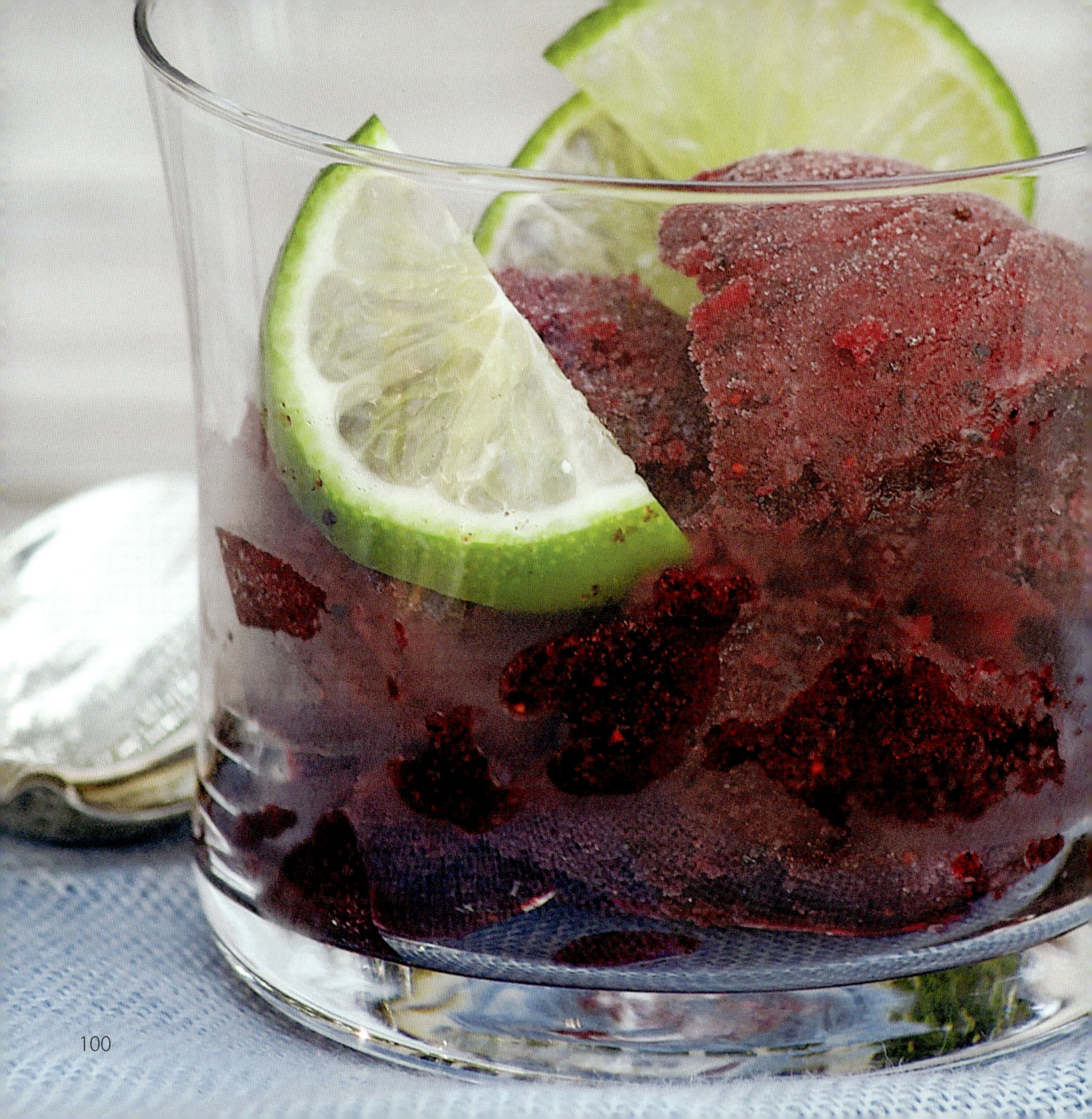

# ice cream, crumbles, and more

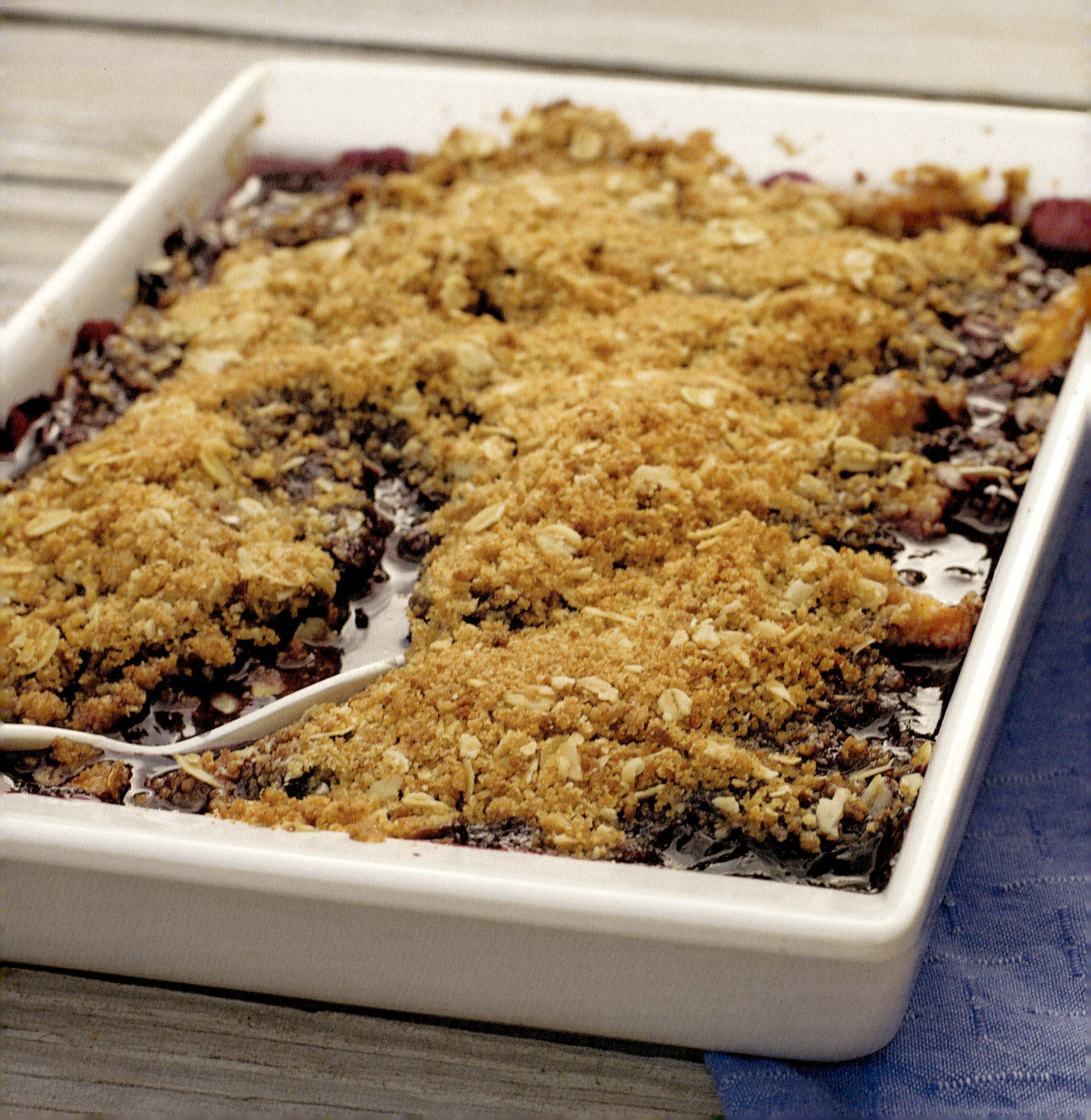

# Peach and Berry Gingersnap Crumble

makes 8 servings

INGREDIENTS

5 cups peeled peaches or nectarines, cut into wedges
2 cups fresh or frozen blueberries
1 quart fresh or frozen strawberries
⅔ cup granulated sugar
2½ tablespoons cornstarch
12 gingersnap cookies (about ½ cup crushed)
⅔ cup all-purpose flour
½ cup firmly packed light brown sugar
½ teaspoon ground cinnamon
½ cup butter, cut into pieces
⅔ cup old-fashioned or quick oats

Stir together peaches, berries, sugar, and cornstarch in a large bowl. Let stand 10 to 15 minutes, tossing occasionally.

Preheat oven to 375°. Butter a 13x9-inch baking dish.

Place gingersnap cookies in a food processor; pulse until fine crumbs form. Add flour, brown sugar, and cinnamon; pulse until well blended. Add butter and pulse until large crumbs form. Add oats and pulse until combined.

Transfer fruit mixture to prepared baking dish. Sprinkle gingersnap topping over fruit.

Cover with aluminum foil and bake 25 minutes. Uncover and continue baking for 15 to 20 minutes or until hot and bubbly.

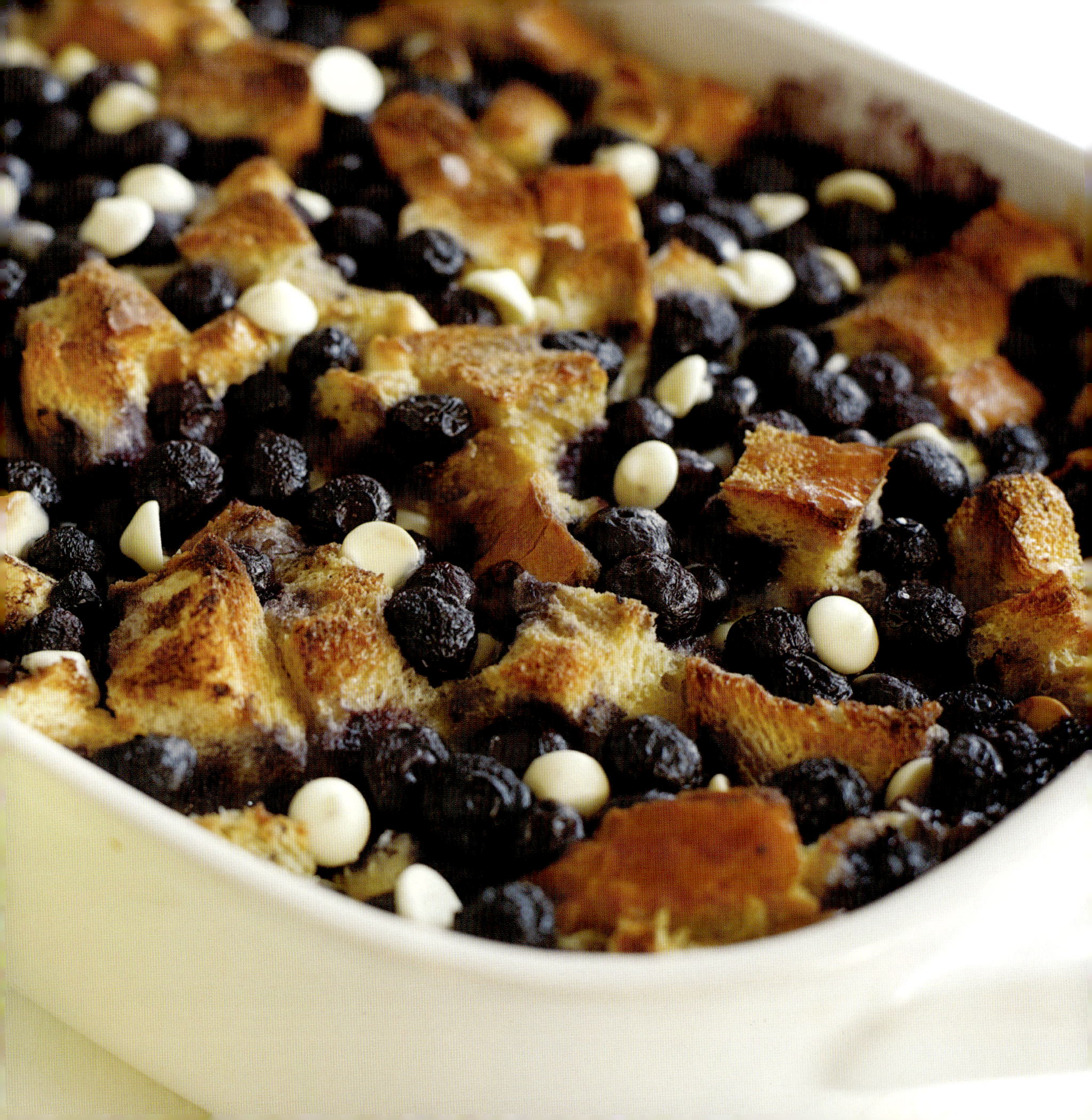

# Blueberry and White Chocolate Bread Pudding

The bread pudding will come out of the oven fairly soft but will firm up as it cools a bit. If desired, substitute 8 ounces cream cheese, cut into small pieces, for the chocolate chips. It's a bit messy to place in the dish with your fingers, but the cream cheese adds another level of creaminess.

makes 8–10 servings

INGREDIENTS

- 4 cups half-and-half or whole milk
- 1 cup granulated sugar
- ¼ cup butter
- 1 (16-ounce) loaf French bread or baguette, stale
- 1 cup chopped white chocolate or white chocolate chips
- 3 cups fresh, frozen, or dried blueberries
- 5 eggs, lightly beaten
- ½ teaspoon cinnamon
- ¼ teaspoon salt
- 1 teaspoon vanilla extract

Preheat oven to 375°. Butter a 13x9-inch baking dish.

Combine half-and-half, sugar, and butter in a medium saucepan over medium heat. Cook, stirring occasionally, until butter melts and sugar dissolves. Set aside; let cool to room temperature.

Cut bread into 1-inch cubes and place in bottom of baking dish. Sprinkle evenly with chocolate and blueberries. Set aside.

Whisk eggs in a small bowl until lightly beaten; whisk into half-and-half mixture. Whisk in cinnamon, salt, and vanilla. Pour half-and-half mixture over bread, chocolate, and blueberries. Press down to soak bread. (Cover and refrigerate overnight, if desired.)

Bake 45 minutes. Let stand 5 minutes. Serve warm.

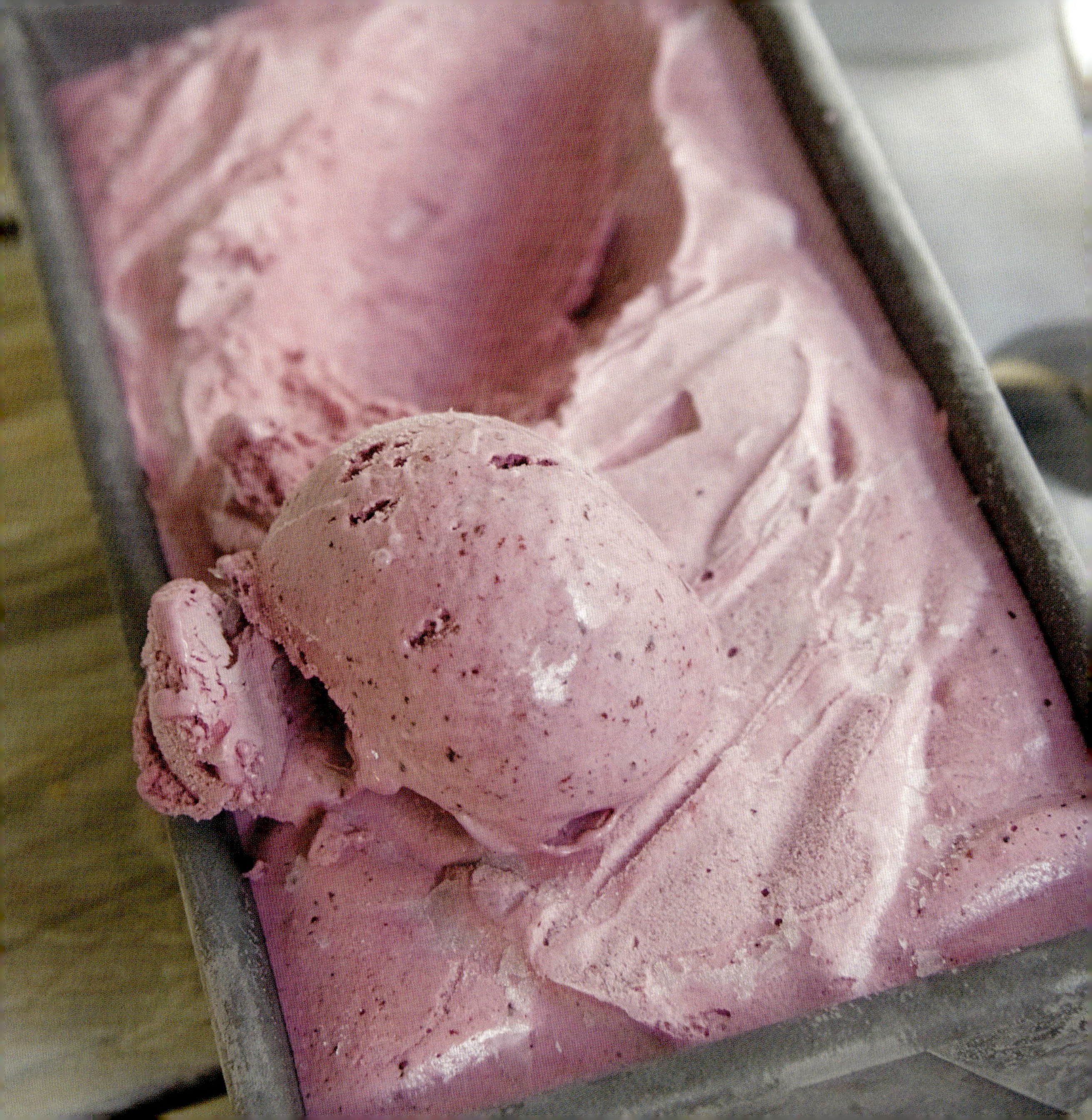

# Blueberry Custard Ice Cream

Seasonal berries and freshly churned ice cream are the best summer combination. This one is old-school, created with a homemade custard base. Be careful when you sample it—it's like the richest cream sauce or a fruity egg nog—you'll be tempted to drink it by the cupful.

makes about 12 cups

INGREDIENTS

4 cups whipping cream
2 cups granulated sugar, divided
1 vanilla bean
8 egg yolks
¼ teaspoon salt
4 cups fresh or frozen blueberries
1 tablespoon fresh lemon juice
2 teaspoons vanilla extract

Fill a large bowl halfway with ice cubes. Pour several cups of water in bowl, about ⅓ full; set aside.

Combine cream and 1½ cups sugar in a heavy nonreactive saucepan over medium heat. Split vanilla bean in half lengthwise; scrape tiny seeds into pan with tip of a sharp knife. Add bean to mixture. Cook, stirring frequently, until sugar dissolves and mixture is hot, but not boiling.

Whisk yolks in a small bowl. Add 1 cup warm cream mixture to eggs, stirring until blended. Add another cup cream mixture, ¼ cup at a time, to egg mixture. Transfer egg mixture back to pan with remaining cream, stirring until blended. Stir in salt. Cook over low heat, stirring constantly, until custard is slightly thickened and reaches 160°. Immediately place pan in bowl of ice water, stirring periodically to cool custard.

Combine remaining ½ cup sugar, blueberries, and lemon juice in a saucepan over medium heat, stirring frequently, 15 minutes, mashing with back of spoon to crush berries.

Strain custard mixture through a fine wire-mesh sieve into a large bowl. Stir in vanilla extract and berry mixture. Cover and chill several hours or overnight. Place in an ice-cream maker and churn according to manufacturer's directions. Freeze.

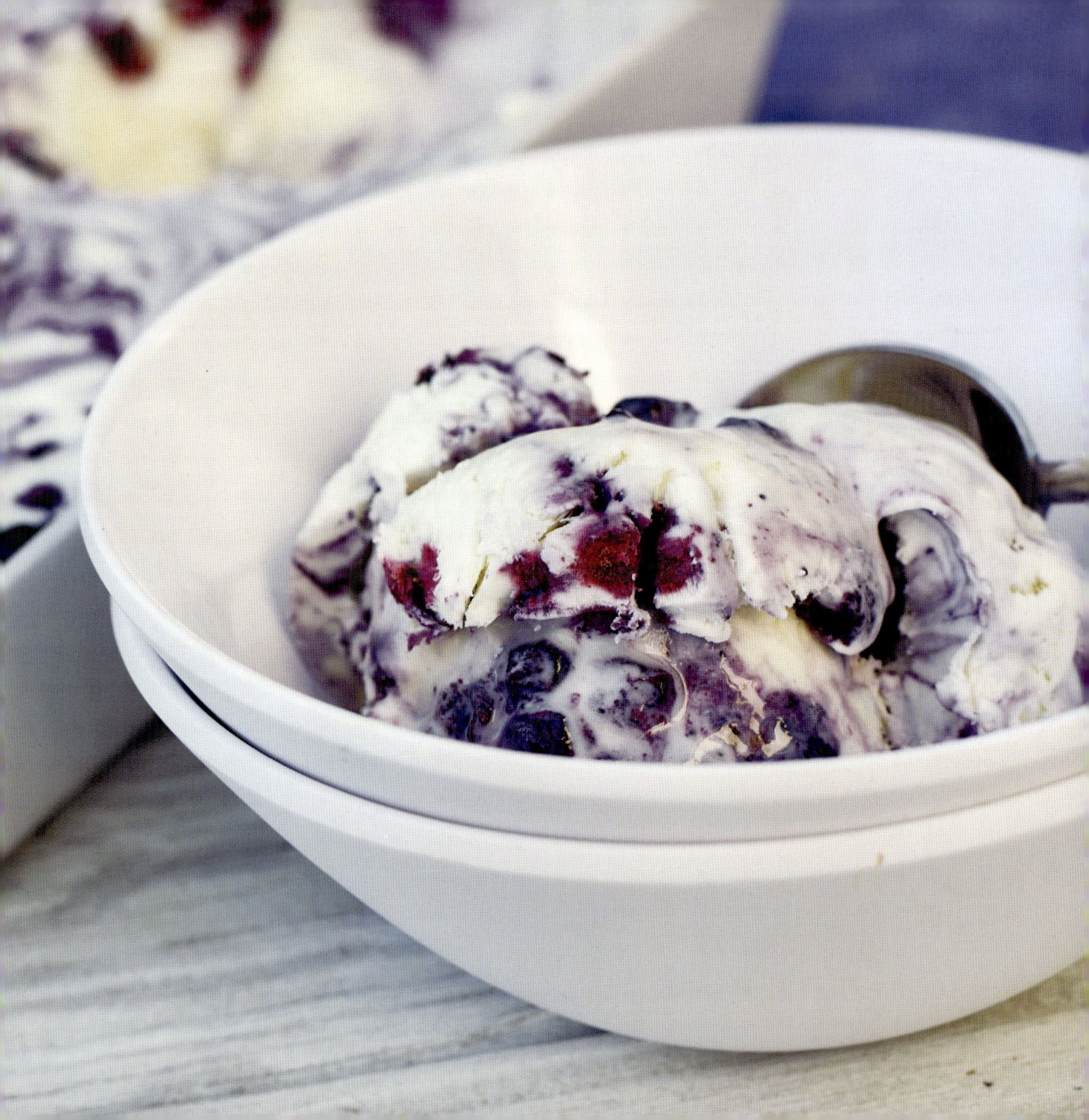

# No-Churn Blueberry Swirl Ice Cream

Try this when you want an easy frozen dessert. Make sure the blueberry sauce is completely cold before mixing together. For very distinct ribbons, put sauce in the freezer until extremely cold but not stiff.

makes 5 cups

## INGREDIENTS

2 cups fresh or frozen blueberries, divided
¼ cup granulated sugar
2 tablespoons water
1 (14-ounce) can sweetened condensed milk
1 teaspoon vanilla extract
2 cups heavy whipping cream

Combine 1 cup blueberries, sugar, and 2 tablespoons water in a small saucepan over medium heat. Cook, stirring constantly, until berries release juice and mixture thickens slightly. Cool blueberry syrup completely; cover and chill until ready to prepare ice cream.

Combine sweetened condensed milk and vanilla in a bowl.

In a separate bowl, beat whipping cream with an electric mixer until soft peaks form. Fold about ½ cup whipped cream into milk mixture to lighten texture. Fold remaining whipped cream into milk mixture. (Do not overmix).

Transfer to a loaf pan, baking dish, or other freezer-safe container. Sprinkle remaining 1 cup blueberries over pan and drizzle blueberry syrup over top. Using a thin spatula, cut into berries and syrup, creating swirls. Cover and freeze several hours or until firm.

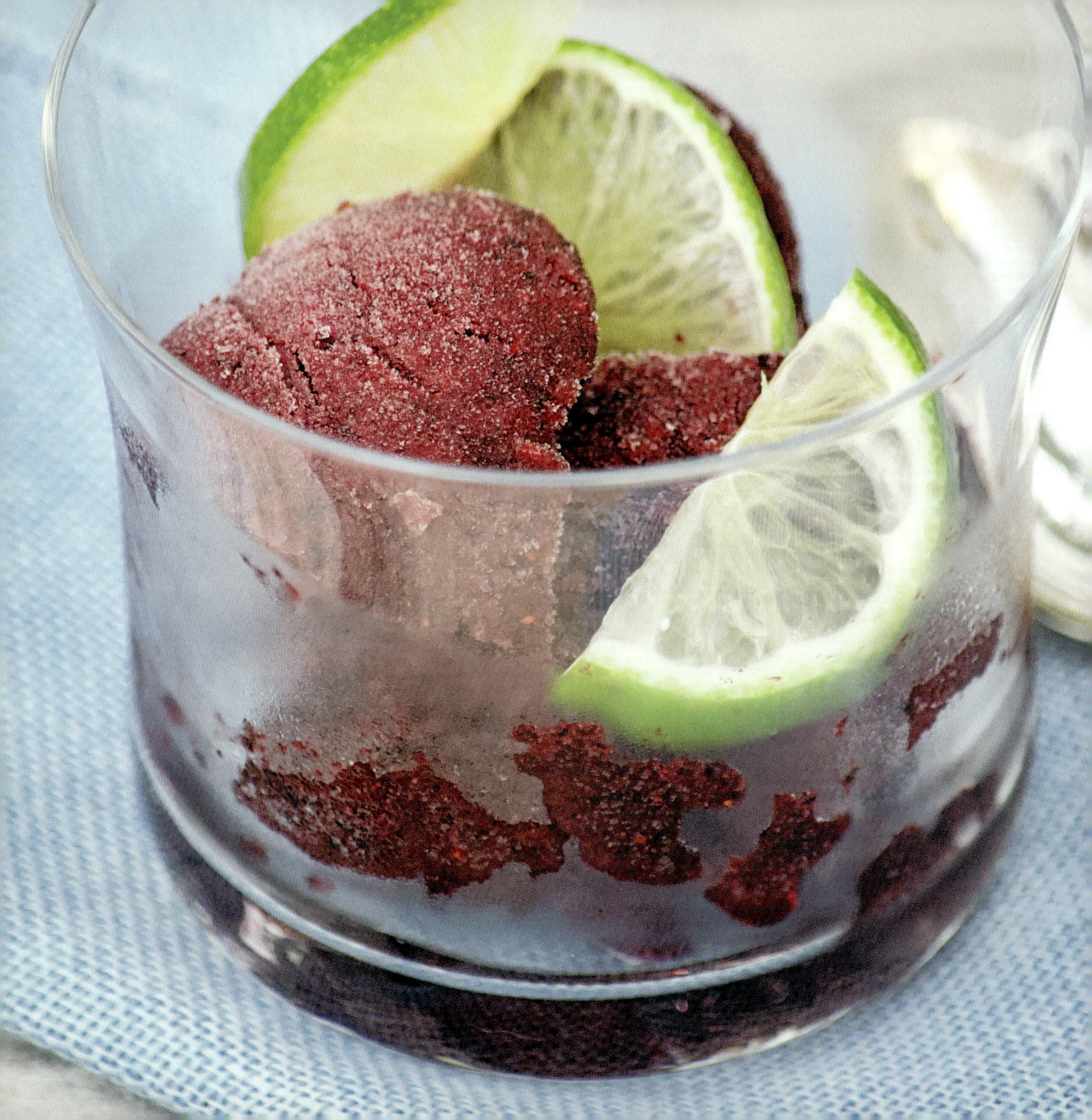

# Blueberry-Lime Sorbet

You can substitute just about any citrus in this recipe—
try lemon, grapefruit, orange, or blood orange for a slightly different flavor.

makes about 4 cups*

INGREDIENTS
1 cup granulated sugar
1 cup water
4 cups fresh or frozen blueberries
2 teaspoons lime zest
½ cup fresh lime juice

GARNISH
lime slices

Combine sugar and 1 cup water in a saucepan over medium heat. Cook, stirring often, until sugar dissolves. Remove from heat and cool completely.

Combine blueberries, zest, and juice in a food processor. Process until smooth. Strain mixture through a wire-mesh sieve, pressing berries with a spoon or spatula. (For a chunky textured dessert, skip straining.)

Stir together sugar mixture and berries; transfer to an 8x8-inch glass baking dish or other freezer-safe container. Freeze, stirring every 20 minutes, until almost firm. Cover and freeze until firm (several hours or overnight). To serve, scrape the surface with an ice-cream scoop or heavy serving fork to lighten texture.

*Blueberry Citrus Pops:* Pour mixture into ice pop molds; freeze until solid. Makes about 1 dozen frozen pops.

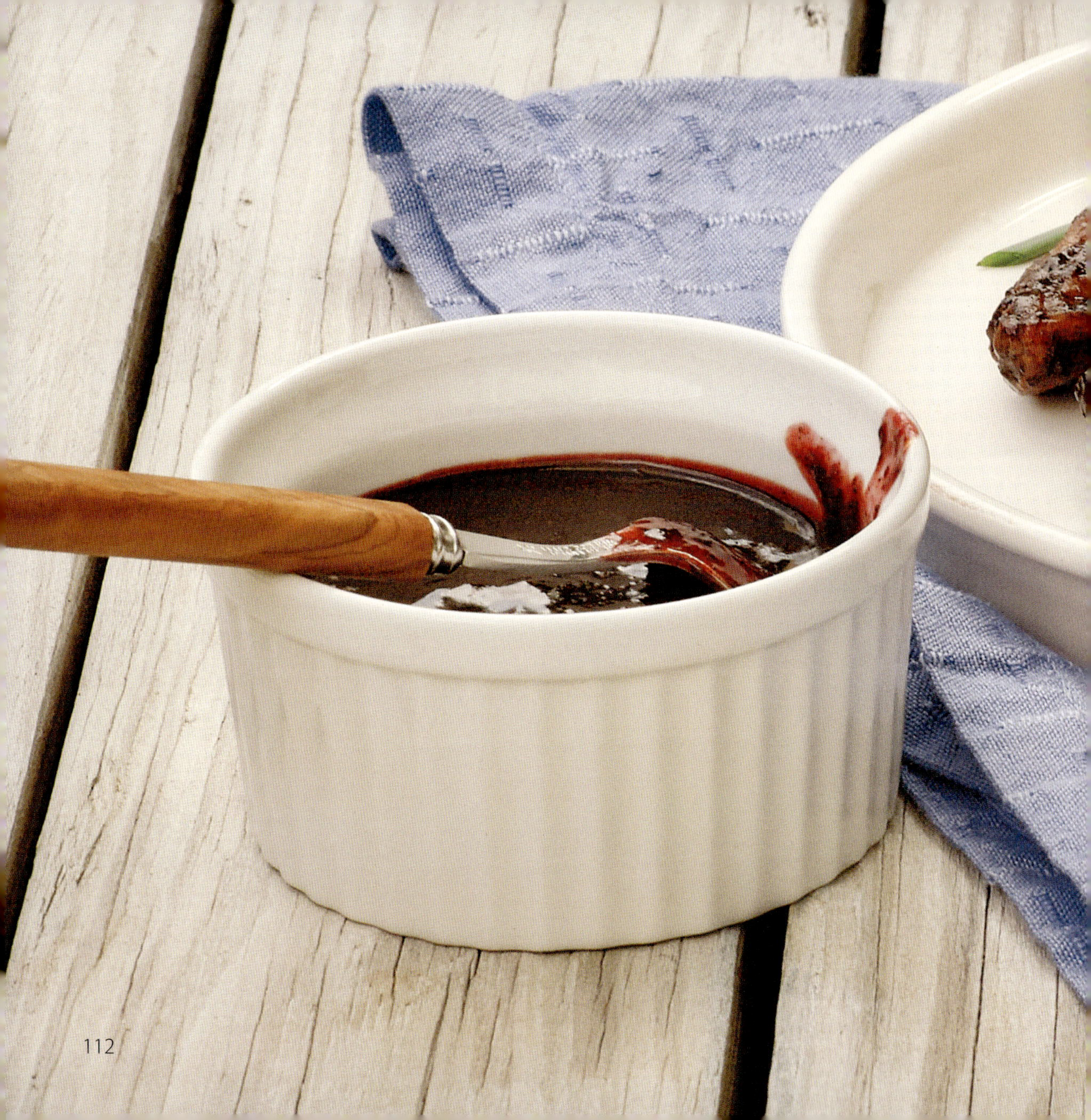

# appetizers, salads, and savory sauces

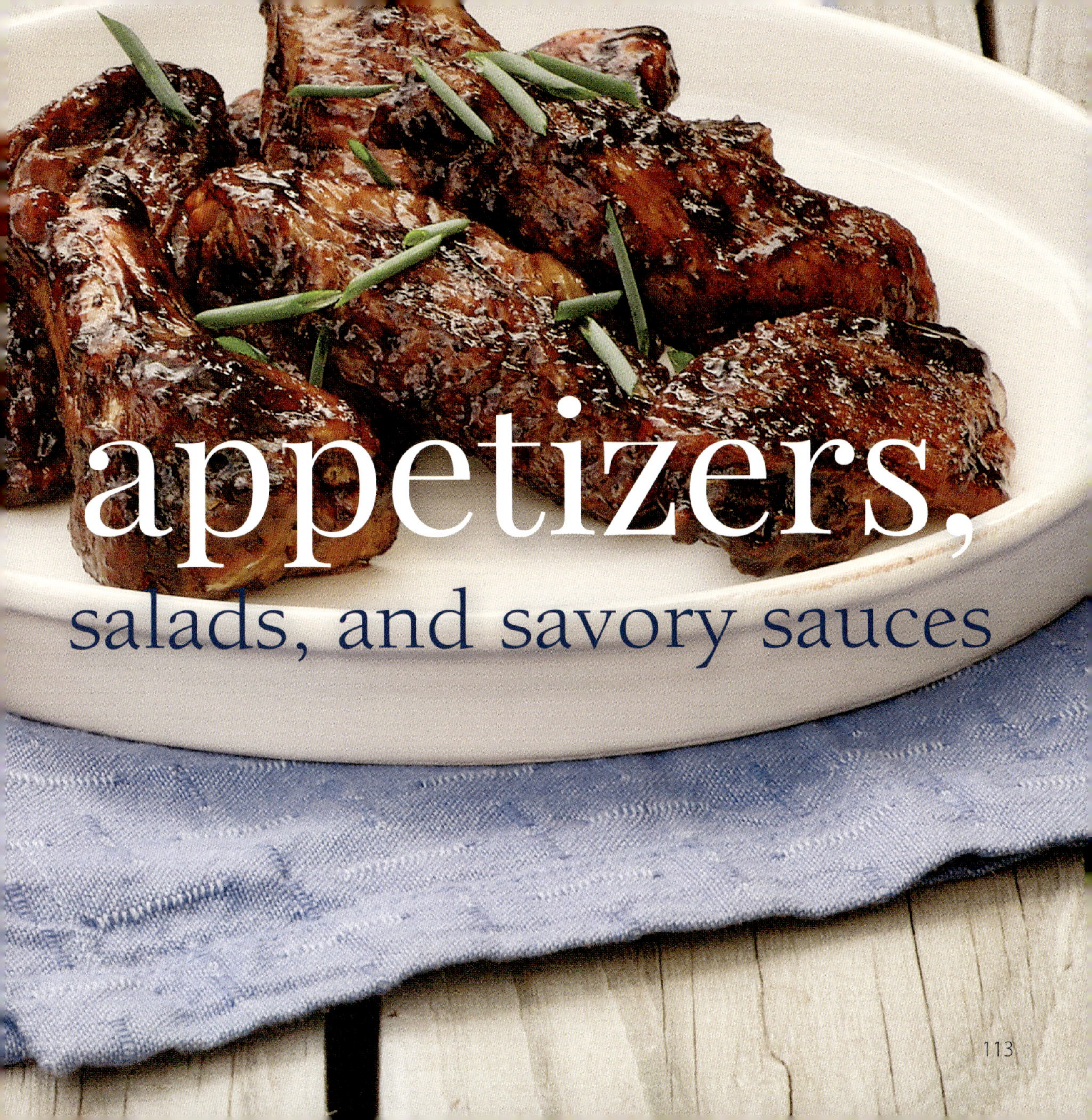

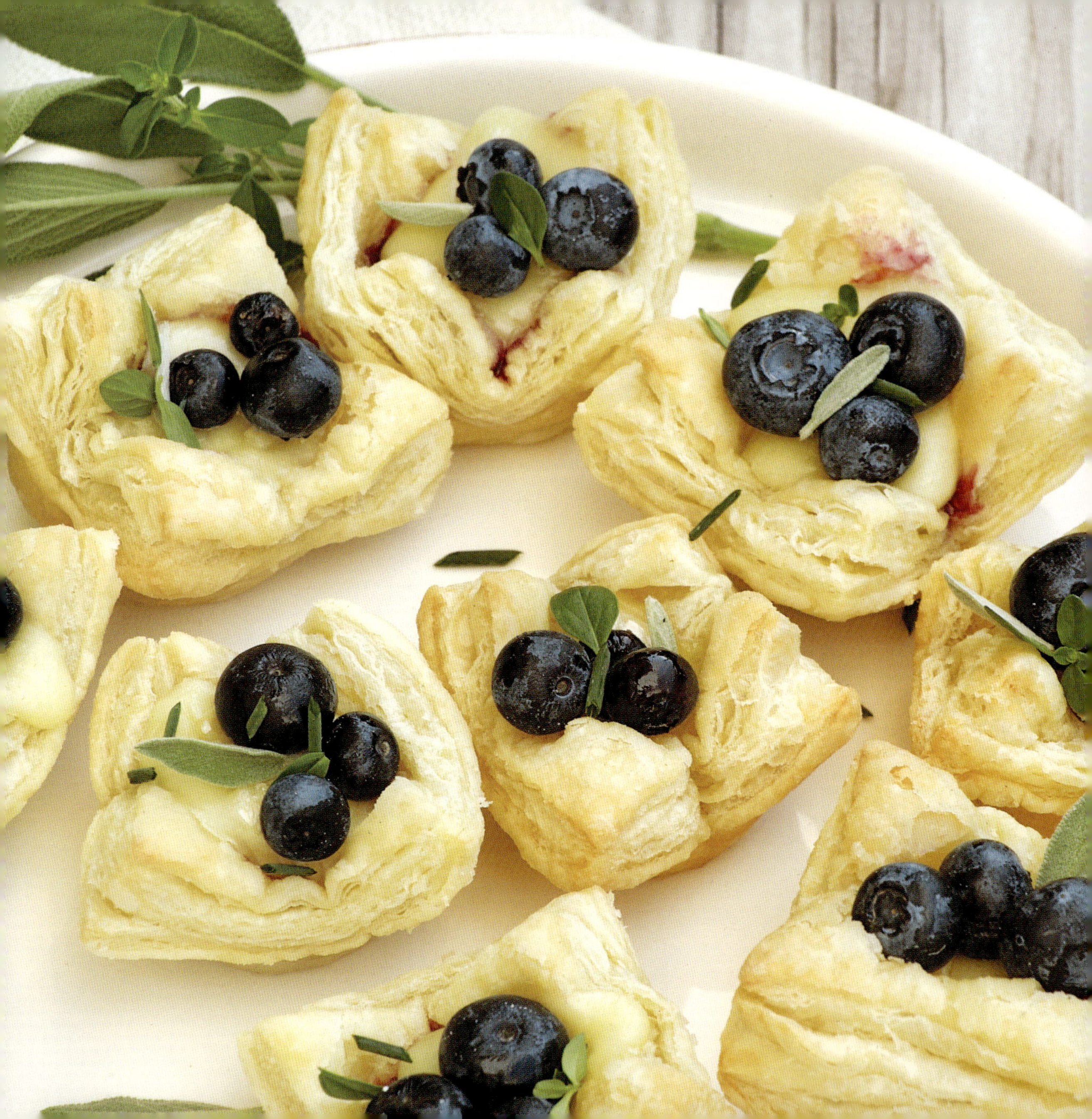

# Blueberry and Brie Tartlets

makes 16

INGREDIENTS

½ (17.3-ounce) package frozen puff pastry sheets, thawed
3 tablespoons Blueberry-Ginger Chutney (page 27) or Simplest Blueberry Jam (page 23)
5 ounces Brie cheese, cut into 16 pieces
½ cup fresh blueberries
1 tablespoon small fresh sage leaves

Preheat oven to 375°.

Roll out 1 pastry sheet on a lightly floured surface. Cut into 16 squares. Fit into cups of 2 mini-muffin pans. Press each center to create a cup. Bake 10 minutes or until golden brown.

Remove from oven and spoon about ½ teaspoon chutney or jam in center of each cup. Top with a piece of Brie. Return to oven and bake 2 more minutes or until cheese slightly melts.

Remove from cups and top with a few blueberries and sage leaves.

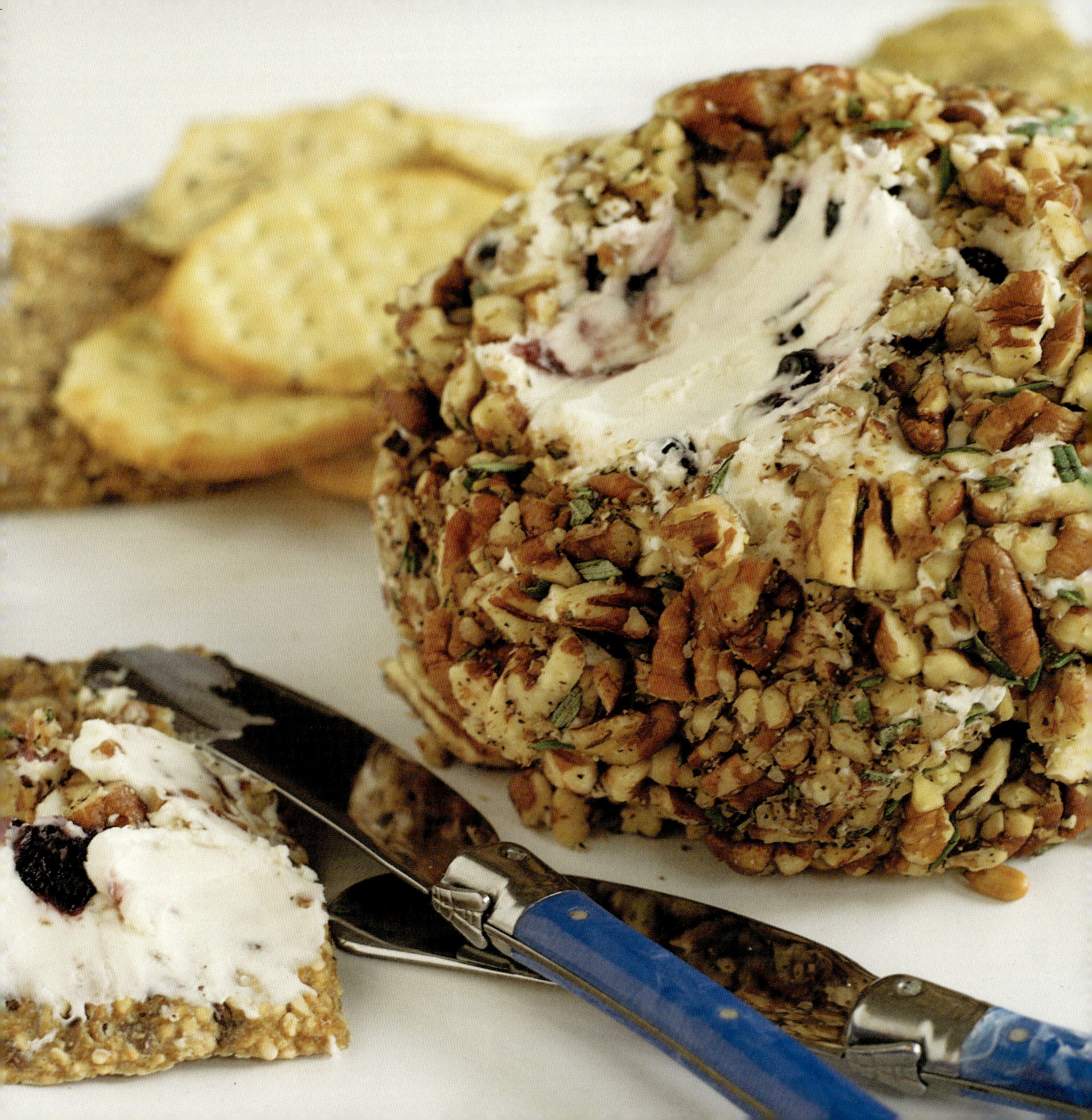

# Blueberry-Pecan Goat Cheese Ball

Don't let the unique combination of ingredients scare you away from this recipe. It's delicious, easy to make ahead, and my most requested recipe from this book! Pressing the herb and nut topping is easier if you place it on a sheet of plastic wrap and roll the cheese ball around.

makes 1⅓ pounds (about 12–20 servings)

INGREDIENTS

- 1 tablespoon butter
- 1 cup chopped pecans
- 1 tablespoon chopped fresh rosemary
- 1 tablespoon chopped fresh parsley
- ½ teaspoon cracked black pepper
- 1 (10- to 11-ounce) log plain goat cheese, softened to room temperature
- 4 ounces cream cheese, softened to room temperature
- 2 tablespoons honey
- ½ cup dried blueberries
- ¼ teaspoon salt

Melt butter in a large skillet over low heat. Add pecans and cook 5 minutes, stirring frequently, until fragrant and lightly toasted. Cool completely and transfer to a small bowl. Stir in rosemary, parsley, and pepper.

In a medium-size bowl, combine goat cheese, cream cheese, honey, blueberries, and salt, stirring until well blended. Form into a ball. Press pecan mixture around ball, and chill until firm.

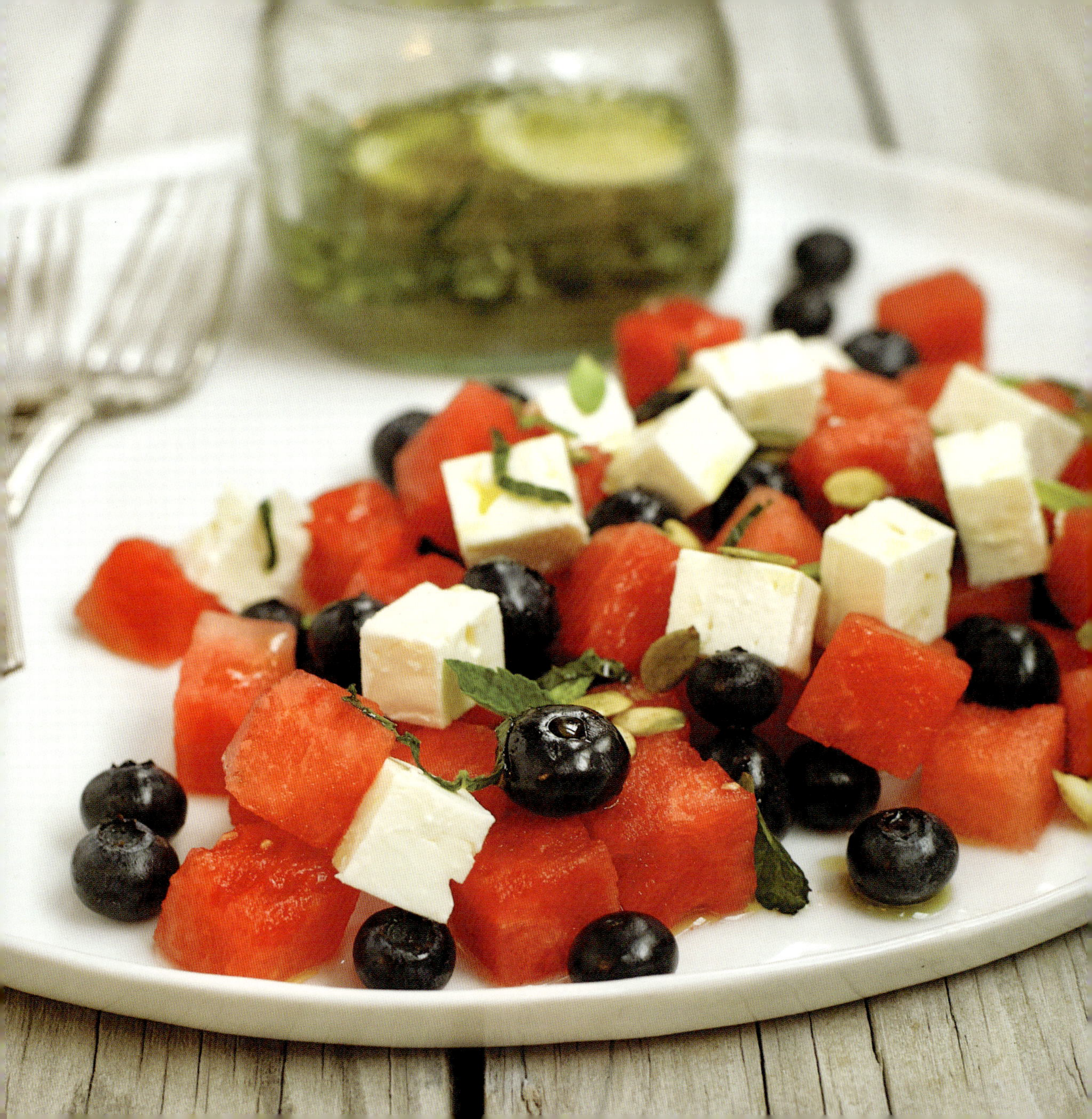

# Watermelon and Berry Salad with Honey-Mint Vinaigrette

Prepare the vinaigrette ahead; chill until you're ready to make the salad.

makes 4 servings

INGREDIENTS

8 cups cubed seedless watermelon
2 cups fresh or frozen blueberries
4 ounces feta cheese, cut into small cubes
¼ cup fresh mint leaves
4 cups fresh arugula, watercress, or mixed greens
Honey-Mint Vinaigrette (recipe at right)
¼ cup roasted pumpkin seeds

Combine melon, blueberries, cheese, and mint in a large serving bowl or individual bowls or plates. Add greens, tossing until combined. Drizzle with desired amount of Honey-Mint Vinaigrette or serve on the side. Sprinkle with pumpkin seeds.

*Honey-Mint Vinaigrette:* Whisk together **¼ cup fresh lemon juice, 1 tablespoon honey, ¼ cup extra-virgin olive oil, ¼ teaspoon salt**, and **⅛ teaspoon freshly ground black pepper.** Add **¼ cup finely chopped fresh mint**, stirring until well blended.

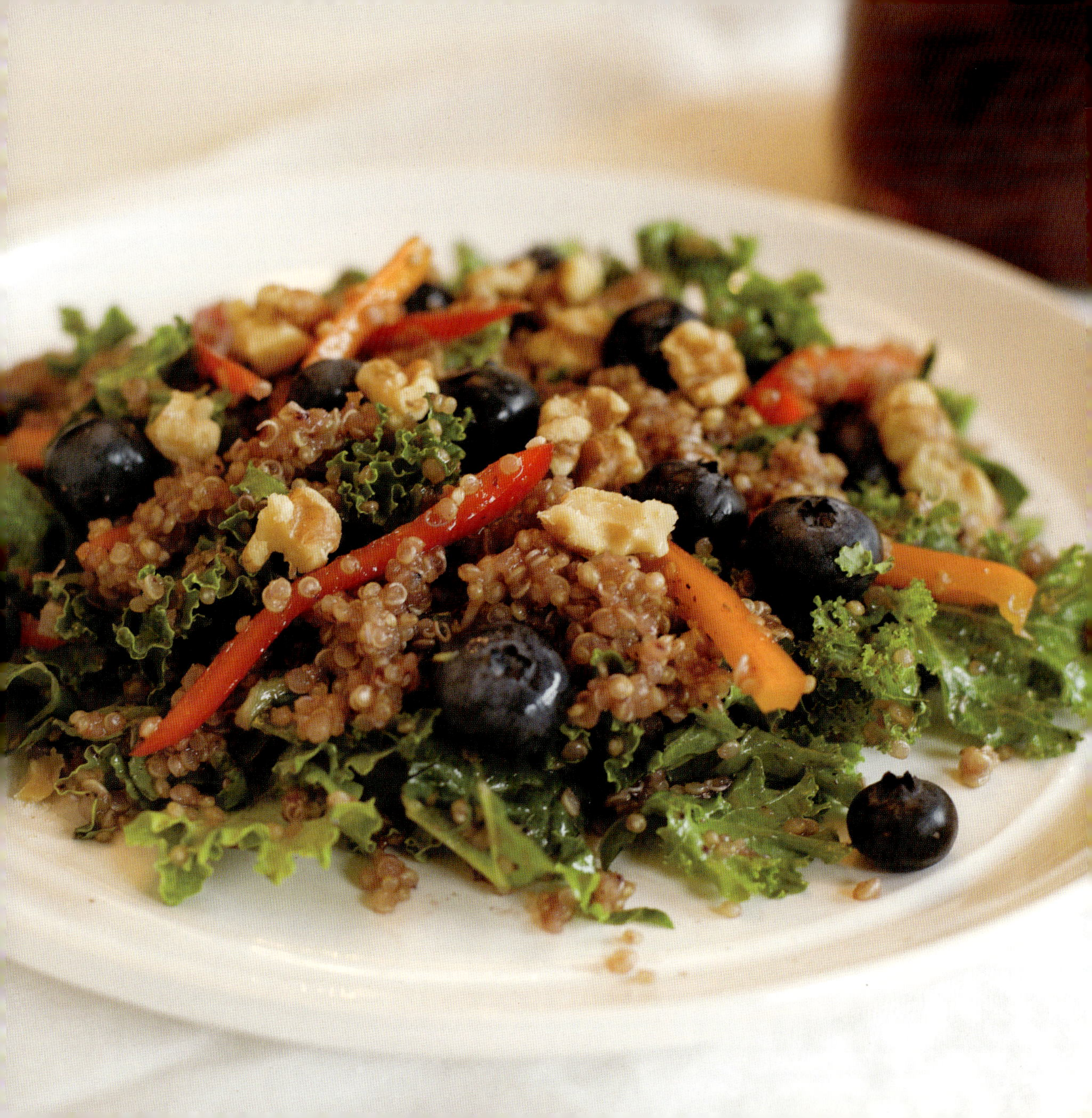

# Crunchy Quinoa Salad with Blueberries and Nuts

This is a superfood salad! Quinoa (pronounced "KEEN-wah") is an ideal food for vegetarian and gluten-free diets because it is considered a complete protein. While it may appear to be a grain, quinoa is related to spinach, chard, and beets, making it completely gluten-free. Prepare the Blueberry Vinaigrette ahead; chill until ready to serve.

makes 4 servings

INGREDIENTS

1¾ cups vegetable broth
1 cup quinoa
1 shallot, minced
1 red, orange, or yellow bell pepper, thinly sliced or chopped
4 cups chopped kale leaves
2 green onions, sliced, or 1 tablespoon chopped fresh chives
Blueberry Vinaigrette (recipe at right)
1 cup fresh or frozen and thawed blueberries
¼ cup toasted walnuts, almonds, or pine nuts, coarsely chopped

Bring 1¾ cups broth to a boil in a medium saucepan; stir in quinoa and shallot. Cover, reduce heat to low, and simmer 20 minutes or until tender. Drain any excess liquid. Transfer to a serving bowl; cover and chill until ready to serve.

Just before serving, toss quinoa with bell pepper, kale, and green onions. Drizzle desired amount of Blueberry Vinaigrette over quinoa mixture and toss again to coat. Sprinkle blueberries and nuts over top.

*Blueberry Vinaigrette:* Combine **1 cup fresh (or frozen and thawed) berries, ⅓ cup apple cider vinegar, 3 tablespoons extra virgin olive oil, 2 teaspoons Dijon mustard, 2 teaspoons honey, ¼ teaspoon salt**, and **¼ teaspoon coarsely ground black pepper** in a blender. Pulse until well blended.

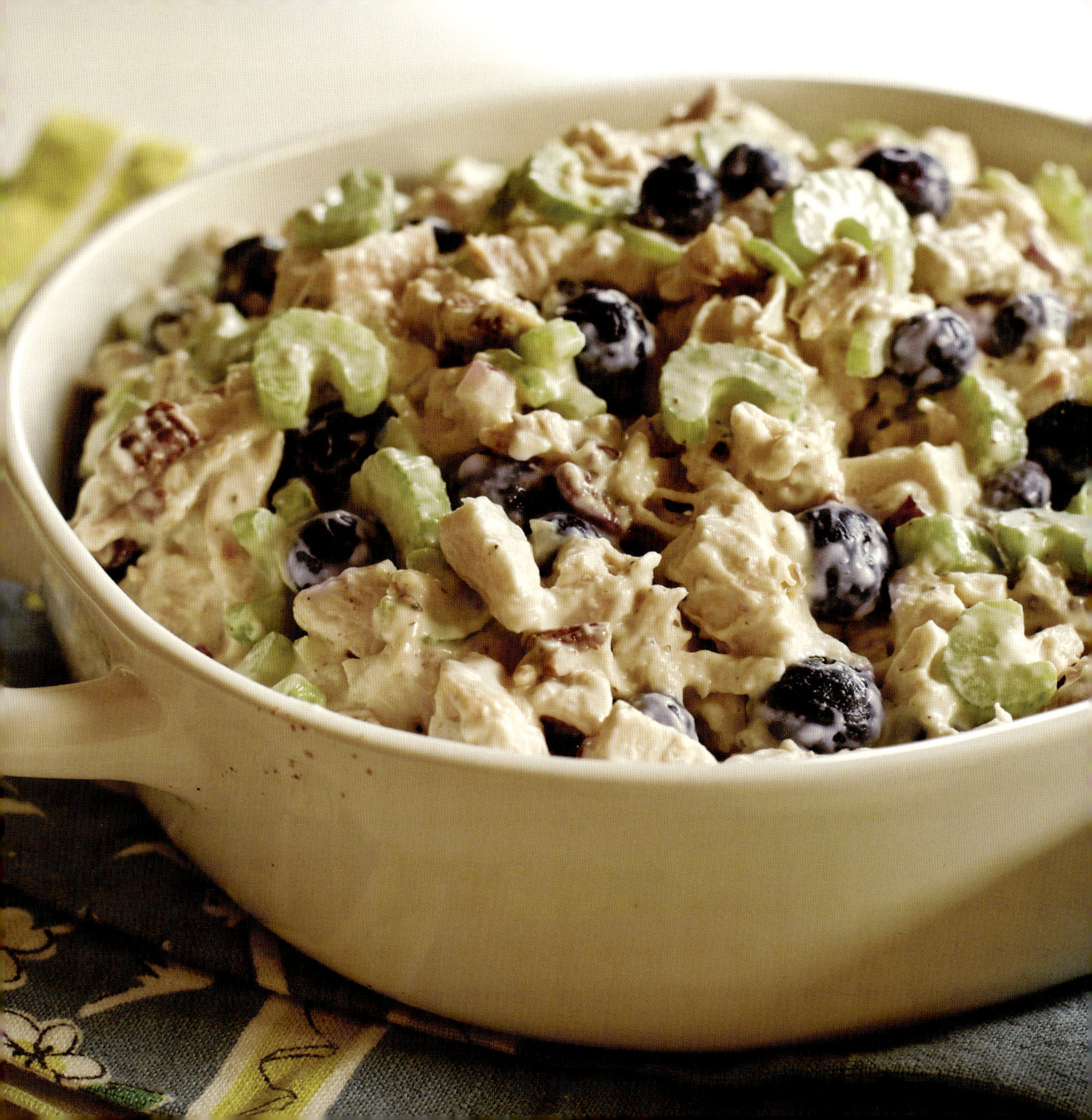

# Honey-Pecan Blueberry Chicken Salad

This recipe makes good use of leftover rotisserie chicken. You probably don't have 4 full cups of leftover chicken—so divide ingredients in half and use what you have. While you can use frozen and thawed blueberries, I prefer dried or fresh because they won't water out and dilute the flavor (or turn the mixture a strange purple-gray color). For a tangy twist, substitute Greek yogurt for mayonnaise and sour cream.

makes 6 cups

INGREDIENTS

- 4 cups chopped or shredded cooked chicken
- 4 small celery stalks, chopped
- 1 cup dried or fresh blueberries
- ½ cup chopped pecans or slivered almonds, toasted
- ¼ cup chopped red onion
- ⅔ cup mayonnaise
- ¼ cup sour cream
- 2 tablespoons honey
- ½ teaspoon salt
- ½ teaspoon coarsely ground black pepper

Combine chicken, celery, blueberries, nuts, and onion in a large bowl.

Stir together mayonnaise, sour cream, honey, salt, and pepper in a small bowl. Stir mayonnaise mixture into chicken mixture.

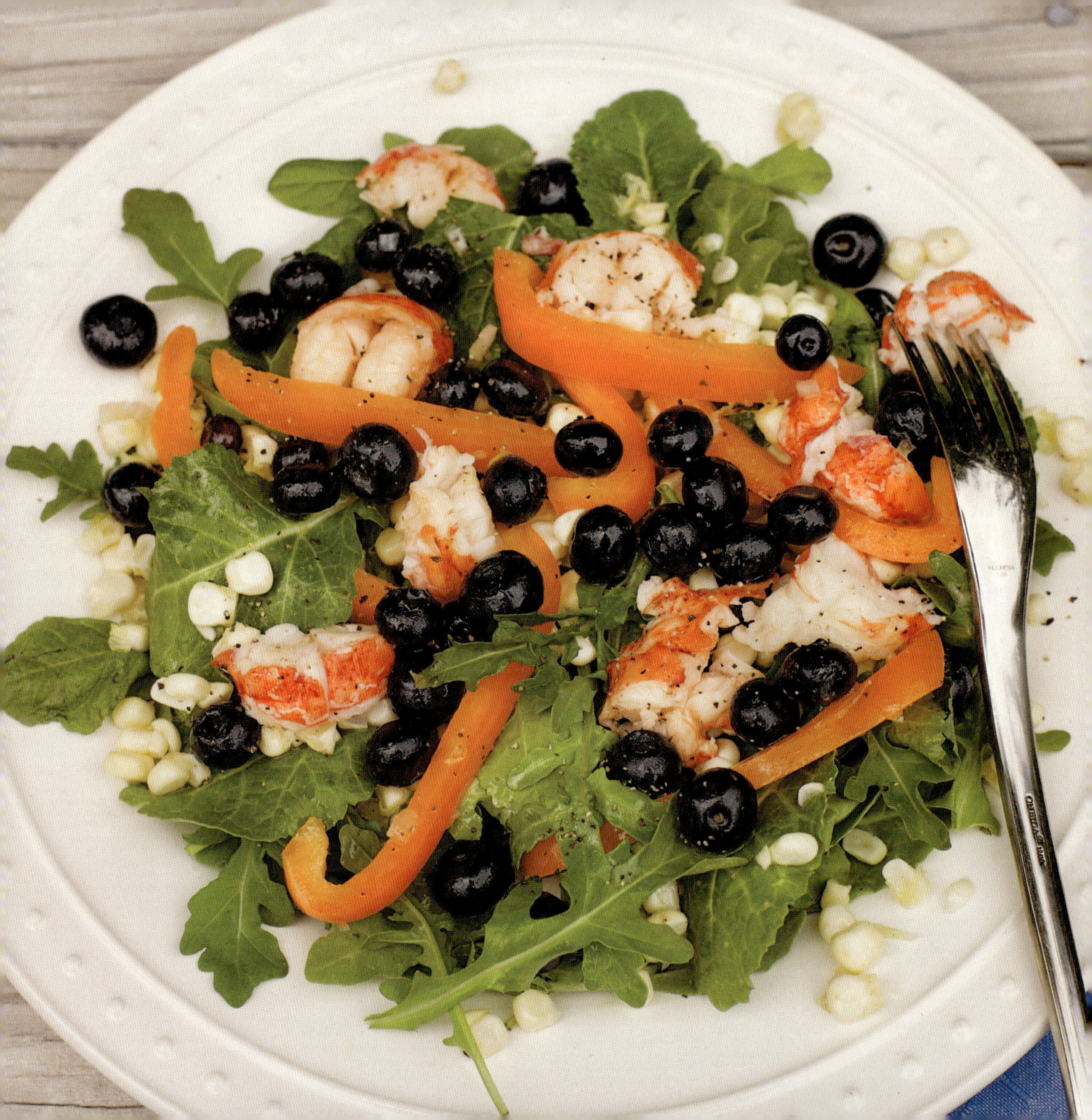

# Blueberry, Lobster, and Corn Salad

Not a fan of lobster? Try shrimp or scallops or even rotisserie chicken. They all pair well with corn and the zesty vinaigrette. When sweet summer corn is plentiful, I cook extra to have for salads and side dishes. It's fine to use frozen in the off-season, and you don't have to cook it—just thaw it in a colander under warm running water. Prepare the vinaigrette ahead and chill until ready to serve.

makes 4 servings

INGREDIENTS

- 3 ears corn (about 2½ cups of kernels)
- ¾–1 pound cooked lobster or 4 cooked lobster tails, cut into bite-size pieces
- 1½ cups fresh blueberries
- 1 yellow, orange, or red bell pepper, cut into slivers
- 1 head Bibb lettuce leaves or 6 cups mixed greens
- Lime Vinaigrette (recipe at right)

Steam, boil, or microwave corn about 3 minutes or until crisp-tender and starchiness is gone. Rinse under cold water to stop the cooking process. Cut kernels off cob.

Arrange lobster, blueberries, bell pepper, and corn over lettuce on one serving platter or individual salad plates. Drizzle with Lime Vinaigrette.

*Lime Vinaigrette:* Whisk together **¼ cup fresh lime juice, 1½ tablespoons sugar, 3 tablespoons rice wine vinegar, 2 tablespoons extra virgin olive oil, 1 minced shallot, ¼ teaspoon salt**, and **¼ teaspoon coarsely ground black pepper** in a small bowl. (Or place in a jar; then cover, and shake to combine).

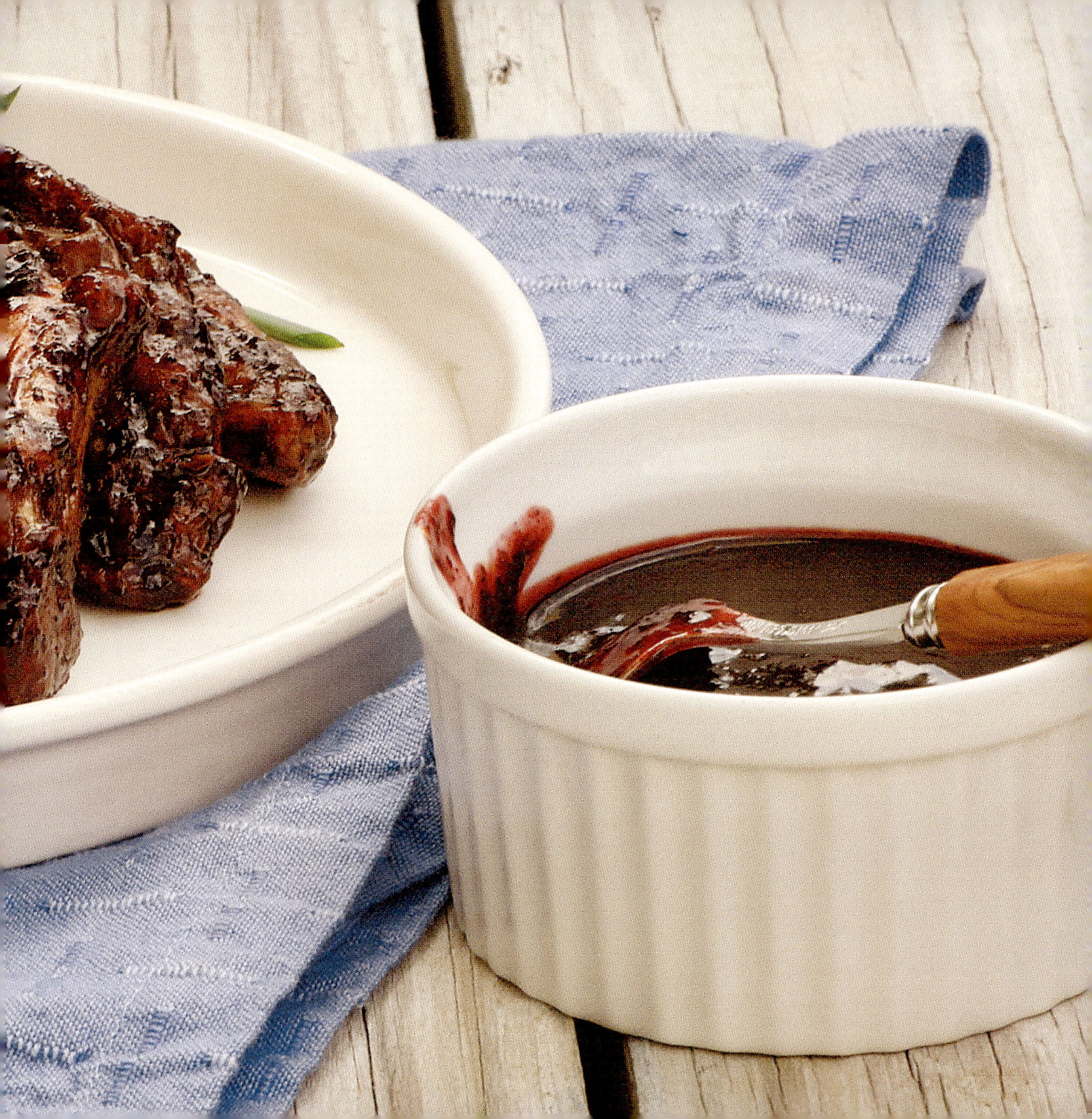

# Blueberry Barbecue Sauce

I've known my friend Lynne Weems Ryan since college, when her roommates were always lucky if she was sharing samples of her amazing dishes. Lynne tosses crispy chicken wings in her spicy-sweet sauce but also recommends serving it with grilled chicken or pork chops.

makes 1½ cups

INGREDIENTS

1 tablespoon safflower oil
1 small red onion, finely chopped
2 garlic cloves, minced
2 jalapeño peppers, seeded and chopped
½ cup brewed coffee
2 cups fresh or frozen blueberries
½ cup ketchup
⅓ cup apple cider vinegar
2 tablespoons firmly packed light brown sugar
2 tablespoons chili powder
1 tablespoon molasses
½ teaspoon salt

Heat oil in a medium saucepan over medium heat. Add onion and cook 2 to 4 minutes, stirring occasionally. Add garlic and jalapeño; cook 1 minute, stirring constantly. Stir in coffee. Bring to a boil; reduce heat, and simmer 5 minutes.

Stir in blueberries, ketchup, apple cider vinegar, brown sugar, chili powder, and molasses. Bring to a boil, reduce heat, and simmer 20 minutes until reduced to about 2 cups and thickened.

Remove from heat and stir in salt. For a smooth texture, grind mixture in a blender or food processor and strain through a fine wire-mesh sieve. Use immediately or cover and refrigerate up to 2 weeks.

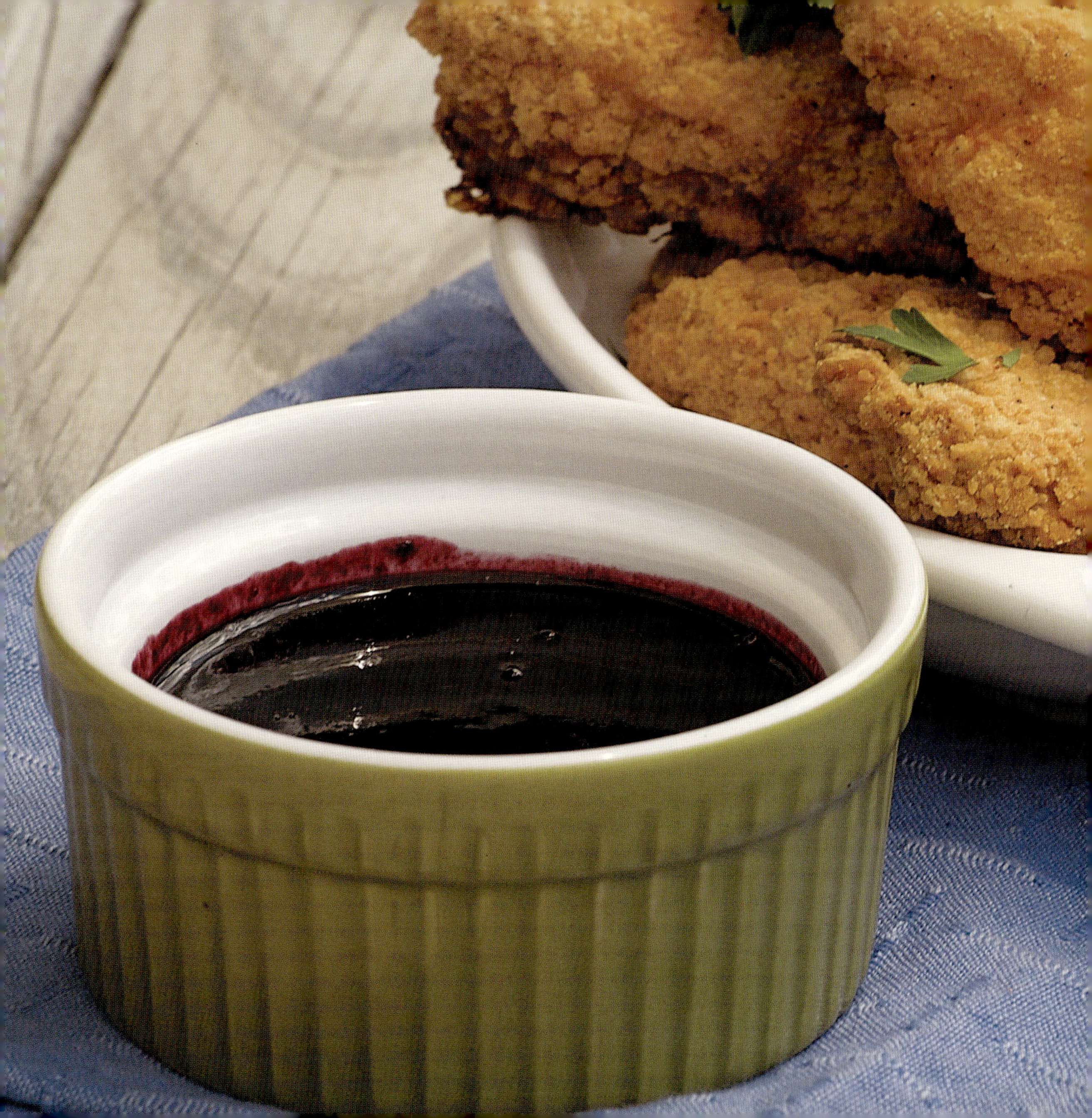

# Blueberry Dipping Sauce for Chicken Fingers

Here's another winning sauce from Lynne. It's easy and delicious—the best combo! Lynne makes the full amount because she's usually feeding a crowd, but I only make half to satisfy my small crew. You can double or halve the recipe—it's foolproof!

makes 1½ cups

INGREDIENTS

1½ cups blueberry jam
1½ tablespoons yellow mustard
1½ tablespoons refrigerated prepared horseradish
1½ teaspoons fresh lemon juice

Combine jam, mustard, horseradish, and lemon juice in a small saucepan over low heat. Cook, stirring occasionally, until warm and well blended.

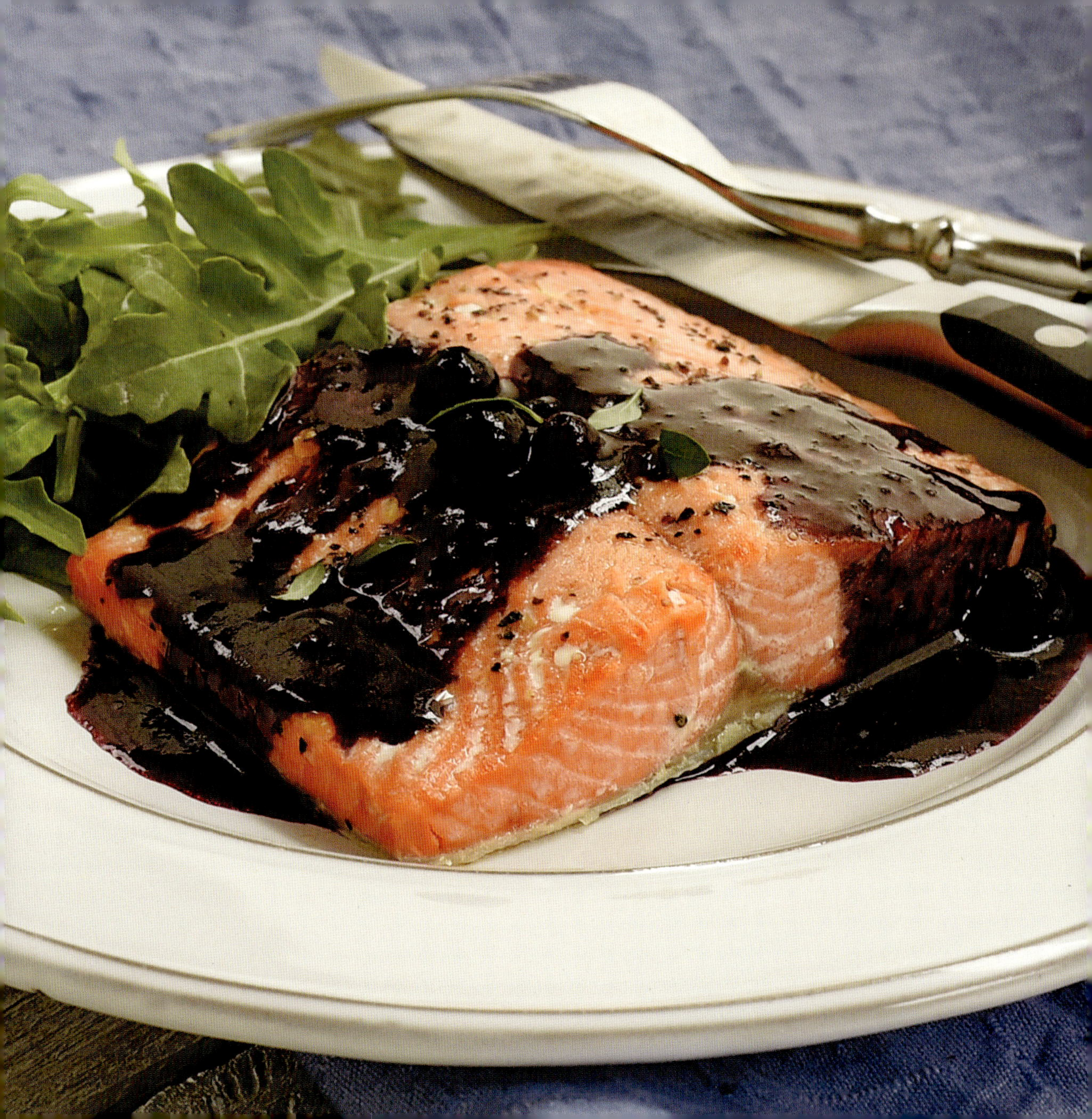

# Blueberry Wine Sauce

The rich, sweet-savory sauce makes an ideal complement to the rich taste of salmon. It's also a nice pairing with grilled game, such as venison or duck, as well as everyday meats such as chicken thighs or quarters.

makes ¾ cup

INGREDIENTS

2 teaspoons extra virgin olive oil
1 large shallot, minced
1 cup fresh or frozen blueberries
½ cup port or red wine
¼ cup red wine vinegar
2 tablespoons granulated sugar
1 teaspoon fresh thyme leaves
½ teaspoon salt
¼ teaspoon coarsely ground black pepper
3 tablespoons butter, cut into pieces

Heat oil in a saucepan over medium heat. Add shallot and cook, stirring constantly, 1 to 2 minutes or until softened.

Stir in blueberries, wine, vinegar, sugar, thyme, salt, and pepper. Cook, stirring frequently, for 10 minutes or until mixture is reduced by half with a syrupy texture. Strain mixture through a fine wire-mesh sieve set over a bowl, pressing mixture with spoon or spatula. Discard pulp.

Remove from heat and whisk in butter, stirring until blended into sauce.

# Index

# About the Author

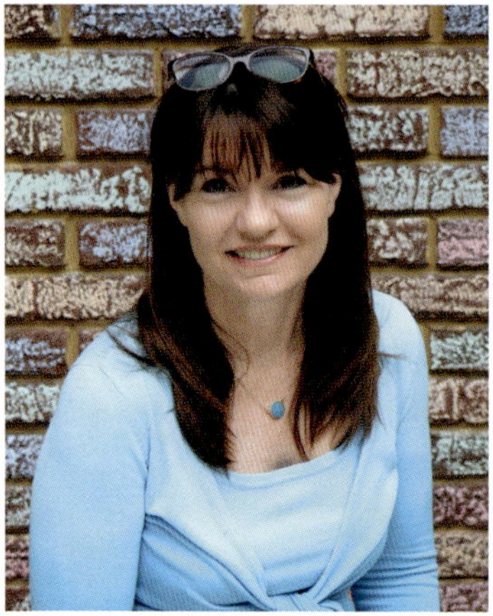

Julia Rutland is a Washington, D.C.-area food writer and recipe developer whose work appears regularly in publications and websites such as Weight Watchers books, *Southern Living* magazine, *Coastal Living* magazine, Myfitnesspal.com, and more. She is the coauthor of *Discover Dinnertime* and a contributor to many other cookbooks and websites.

Julia lives in Purcellville, Virginia, with her husband, two daughters, a cat, a couple of dogs, too many chickens, and whatever animals decide to adopt them.

Visit Julia online at www.juliarutland.com